MARKET ORIENTATION, CORPORATE CULTURE AND BUSINESS PERFORMANCE

Market Orientation, Corporate Culture and Business Performance

SATYENDRA SINGH
University of Winnipeg, Canada

ASHGATE

Published by
Ashgate Publishing Limited
Gower House
Croft Road
Aldershot
Hants GU11 3HR
England

Ashgate Publishing Company
Suite 420
101 Cherry Street
Burlington, VT 05401-4405
USA

Ashgate website: http://www.ashgate.com

British Library Cataloguing in Publication Data
Singh, Satyendra
Market orientation, corporate culture and business performance
1. Relationship marketing 2. Corporate culture 3. Machine-tool industry - Marketing
I. Title
658.8'12

Library of Congress Cataloging-in-Publication Data
Singh, Satyendra, 1966-
Market orientation, corporate culture and business performance / Satyendra Singh.
p. cm.
Includes bibliographical references and index.
ISBN 0-7546-3578-3
1. Machine-tool industry--Great Britain--Management. 2. Machine-tools--Great Britain--Marketing. I. Title.

HD9703.G72S53 2004
621.9'02'0688--dc22

2004002441

ISBN 0 7546 3578 3

Printed and bound in Great Britain by
Athenaeum Press Ltd., Gateshead, Tyne & Wear

Contents

List of Figures

List of Tables

List of Abbreviations

AMTRI	Advanced Manufacturing Technology Research Institute
ANOVA	Analysis of Variance
CAD	Computer Aided Design
CAM	Computer Aided Manufacturing
CNC	Computer Numeric Control
DNC	Direct Numeric Control
DV	Dependent Variable
FAME	Financial Analysis Made Easy
FEA	Finite Element Analysis
FMS	Flexible Manufacturing System
GUI	Graphic User Interface
HMV	Hypothesized Moderator Variable
ISO	International Standard Organisation
LISREL	Linear Structural Relation Equation
MIT	Multiplicative Interaction Term
MO-INDEX	Market Orientation Index
MRA	Moderated Regression Analysis
MSI	Marketing Science Institute, USA
MTI	Machine Tool Industry
MTTA	Machine Tool Technology Association
NC	Numeric Control
NPS	New Product Success
OP	Overall Performance
PCBN	Polycrystalline Cubic Boron Nitride
PCD	Polycrystalline Diamond
R&D	Research and Development
ROA	Return on Assets
ROCE	Return on Capital Employed
ROI	Return on Investments
SAS	Service after Sales
SBU	Strategic Business Unit
SD	Standard Deviation
SG	Sales Growth
SGA	Subgroup Analysis
SIC 3541	Standard Industrial Classification Code, Metal-Cutting
SIC 3542	Standard Industrial Classification Code, Metal-Forming
SME	Small and Medium-Sized Enterprizes
SPSS	Statistical Package for Social Scientist
TQM	Total Quality Management
VIF	Variation Inflation Factor

Acknowledgements

I gratefully thank the following professors and professionals for their guidance and suggestions in the preparation of the book: Kwaku Appiah-Adu, Government of Ghana, Ghana; Gin Chong, Southampton Institute, UK; Tony Hope, Southampton Institute, UK; Peter Jennings, Southampton University, UK; John Latham, Southampton Institute, UK; Geoff Noon, Machine Tool Technological Association, UK; Ashok Ranchhod, Southampton Institute, UK; Caroline Tynan, Nottingham University, UK; Tim Wheeler, University Chester College, UK. My thanks are also due to Alex Martinez and Angela Davis for checking typographical errors in the book. Finally, it was my good fortune to work with the editors: Brendan George and Mary Savigar at Ashgate Publishing Ltd., England; thanks to them as well.

Preface

This book examines the link between market orientation and business performance in a sector-based environment and detects if the relationship between market orientation and business performance is moderated by environmental factors. By conducting the research in a differentiated sector (machine tools), the effects of environmental factors such as market turbulence, technological turbulence, competitive intensity, market growth, buyer power, seller concentration, among others had the same controlling effect for all the players in the sector. Further, controlling of such factors is particularly relevant when the industry is characterised by a large number of small companies manufacturing a wide variety of types and size of products.

Although most studies suggest that market orientation leads to better business performance, this notion holds true (in the case of machine tool manufacturing firms based in the UK) only if Sales Growth, Market Share, New Product Success, Return on Investment, Customer Retention, and Global Presence of companies are taken as business performance indicators. Market Orientation does not have a significant positive impact on the Overall Performance of machine tool manufacturing companies. Further analysis of the data suggests that relationship between Market Orientation and Business Performance is stronger when Market and Technological Turbulences are low, and Competitive Intensity is high. It is argued that managers do not practice customer orientation and competitor orientation at the same time till they have achieved an optimum level of business performance. However, in order to increase business performance further, managers do practice both customer and competitor orientation simultaneously if they have crossed the optimum level of business performance.

In addition, an operational machine tool industry-specific market orientation scale has been developed which has avoided conventional focus on single-authored measure of market orientation, and rather adopted a multifaceted view of the concept. Similarly, the performance measures are based on a multidimensional view of financial and non-financial indicators. A new performance indicator, global presence, has been utilized in the study to examine the relationship between market orientation and the global presence of companies.

Managers can use the market orientation scale to measure the level of market orientation in their own firm and examine if they should commit resources - in terms of time and money - to be market-oriented. Managers will also benefit from the book by understanding the circumstances under which market orientation-business performance relationship may not be significantly positive.

Chapter 1

Introduction

Objectives

The environments of most businesses are currently characterised by increasing competition and environmental turbulences. Most firms have had to find ways of dealing with this stark reality or face the possibility of extinction. As a consequence of the increasing efforts by managers to develop a competitive edge in their respective business sectors, the management literature is replete with conceptual propositions on sound business practices and strategies for success in today's competitive marketplace (Day, 1994). In this context, the market orientation concept has received considerable attention from practitioners as well as academic researchers. Marketing is regarded as a driving force for business strategies and operations for better performance. In recent years, there has been a series of debate over the impact of market orientation on business performance. This study has been designed to contribute to the debate.

Although earlier research on market orientation tended to focus on cross sectional studies to contribute to theory building and to examine the universal importance of the concept, recent empirical efforts have tended to be industry-specific (Morgan and Morgan, 1991 in the consulting engineering profession; Liu, 1996 in manufacturing; Singh, 1998 in the machine tool industry; Chee and Peng, 1996 in the housing market, Soehadi et al., 2001 in retail sector; Gainer and Padanyi, 2002 in nonprofit organizations; Tay and Morgan, 2002 in chartered surveying industry, and others). One particular feature of the literature is that most studies have focused on the relationship between market orientation and business performance, with the majority of studies reporting a positive association between the two variables. Clearly, findings from these studies on the consequences of a market orientation are important, because they can provide managers with the knowledge associated with factors required for developing a market-oriented concept. Therefore, the overall aim of the project is to develop a comprehensive understanding of the subject by researching into the underlying theoretical framework of market orientation-business performance relationship, and to test the hypotheses empirically relating to these variables in the context of machine tool industry.

Further, it investigates the issues relating to market orientation, such as, can firms achieve a competitive advantage by implementing the concept of market orientation; does the implementation of this concept have a positive impact on business performance, as measured by efficiency-, effectiveness- and adaptability-based performance indicators; and, is market orientation-business performance relationship moderated by market forces, such as competitive intensity, market turbulence, and technology turbulence. The introduction of the book begins by identifying the underlying dimensions, representing market orientation, and testing the relationships between each of the underlying dimensions of market orientation and the business performance.

More specifically, the objectives of the study are: to redevelop a comprehensive market orientation scale that delineates the theory of market orientation; to discover the significant underlying dimensions of the market orientation construct and examine the impact of these dimensions on business performance; to evaluate the extent to which the machine tool industry is market oriented and its effect on the performance of the companies, and to identify potential moderator variables, which could influence the market orientation-business performance relationship; and to investigate if corporate culture, represented by the four dimensions - Market culture, Adhocracy culture, Clan culture and Hierarchical culture - is an antecedent to the market orientation.

Rationale for the Study

A review of the literature reveals that the majority of the studies on market orientation appear to be cross-sectional in nature (Houston, 1986; Shapiro, 1988; Webster, 1988; Deshpande and Webster, 1989; Kohli and Jaworski, 1990; Narver and Slater, 1990; Norburn et al., 1990; Greenley, 1995; and Liu, 1996). This undifferentiated sectorial study creates its own problems in understanding the effect of the business environment. By carrying out this research within the machine tool industry (or in any particular industry), some of the environmental variables such as market growth, buyer power, seller concentration, competitive intensity and technology among others can have the same controlling effect for all the players in the sector. There is a paucity of empirical research in this sector. Therefore, this study seeks to examine the market orientation-performance link in the machine tool industry. This sector is acknowledged as an indicator for the health of the manufacturing industry, as many industries rely on the machine tool industry to supply innovative manufacturing equipments.

Despite the growing interest of researchers in the area of market orientation and its measurement, it appears that no systematic research has examined the role and impact of market orientation in the machine tool industry. In this study, an attempt is made to reconcile the three dominant market orientation scales, i.e. Narver and Slater (1990), Jaworski and Kohli (1993), and Deng and Dart (1994) to redevelop a more comprehensive domain of the scale. Applying this scale within the machine tool industry and investigating the operational modifications should help develop an effective market orientation scale.

Definition of Market Orientation

With regard to the definition of market orientation, the study examines critically the definitions given by Kohli and Jaworski (1990), Narver and Slater (1990) and Deng and Dart (1994). The definitions given by these researchers are as follows:

Kohli and Jaworski (1990) define market orientation as the combination of three broad business activities: the generation of market intelligence, the dissemination of this intelligence and organization-wide responsiveness to it. The Narver and Slater (1990) definition of market orientation complement the definition of Kohli and Jaworski (1990) with three behavioral components – customer orientation, competitor orientation, inter-functional coordination – and two decision criteria – long-term focus and profit objective. Deng and Dart (1994) have combined the definition of Kohli and Jaworski (1990) and Narver and Slater (1994) which emphasises on the implementation of a particular business philosophy, the marketing concept, and profit, considering it to be the main objective of a firm because profit orientation is an inherent practice in the day-to-day operation of most successful businesses.

As there is no consensus as to what market orientation is, I have defined, based on the definitions mentioned above and on the interviews with business managers, the following definition of market orientation to serve a starting point for the study.

Market orientation is a set of organization-wide activities coordinated in such a way that derives customer satisfaction through superior performance of products while still being competitive in the marketplace. (Singh, 1998)

Why the Machine Tool Industry?

In the last two decades there has been a shift in the dominance of individual countries within the sector. For example, the British machine tool industry has declined dramatically whereas some European manufacturers have either declined slowly or have maintained their position. The relative decline of the British machine tool industry has been a subject of debate for generations.

The technology of the machine tool industry is vital for the creation of all modern products. Machine tools are used by all sectors of manufacturing including aerospace, automotive, defence, railways, construction equipment, agricultural machinery, consumer durable and many other applications. Because machine tool manufacturers typically sell their products worldwide, a weak domestic machine-tool industry means that manufacturers risk losing access to the foreign markets. This makes the machine tool sector a worthy of study as it has a significant impact on the economy of the nation. This research project is based on an in-depth study of the Machine Tool sector in the United Kingdom. This particular strand of research was carried out as a result of the call to examine whether the market orientation-business performance relationship was present in high-tech industries (Narver and Slater 1990).

Further, it is also a sector, which until now has been studied solely from an economics or engineering point of view. The machine tool sector (SIC 3541 and SIC 3542) was selected for the study as it acts as a barometer of the manufacturing industry as a whole. As the industry's sales volume can be taken as a leading indicator of economic growth or slow down (Balkrishnan, 1996), studying the industry from a marketing viewpoint would enhance our understanding of the market forces under which the sector might be more profitable.

Contribution to the Body of Knowledge

The study contributes to the existing literature on market orientation in a number of ways.

First, from a theoretical point of view an operational machine tool industry-specific market orientation scale has been redeveloped. This study has avoided the conventional focus on single authored scale of market orientation, and rather adopted a multifaceted view of the concept. This makes this scale more representative of market-oriented activities. Similarly, the performance measures are based on a multidimensional view of financial and non-financial indicators. A new performance indicator, global presence, has been used in this study to examine the relationship between the market orientation and companies' global presence.

Second, from a managerial point of view, it considers the degree to which market orientation factors are related to business performance. This study also identifies the circumstances under which a market orientation-business performance relationship may not be significantly positive. Therefore, managers can use the market orientation scale to measure the level of market orientation of their own firms, and examine if they should commit resources in terms of time and money to increase the market orientation of their companies.

Third, empirically, it examines the characteristics of underlying factors of market orientation in the UK machine tool industry. It provides managers with insight as to the selection of significant factors applicable in the industry.

Finally, from a corporate culture point of view, this study offers empirical insight into the relationship between corporate culture and market orientation.

Organization of the Book

The theoretical framework and the organization of the study are illustrated in Figures 1.1 and 1.2, respectively. These figures are essential in understanding the nature and significance of the study. The book contains eight chapters. To offer a summary of the substance of the book, a brief description of each chapter is presented in the following section.

Chapter two begins with the discussion about prevailing wisdom that having a market orientation is critical to superior performance and long-term success of businesses, particularly in an environment which is characterized as being competitive and volatile. A review of literature offers a mixed support for the premise, which leads us to examine the environment in which the relationship between market orientation and business is significantly positive in the context of the machine tool industry. Given the inconsistency in the findings among previous studies based on market orientation, theoretical and conceptual foundations for the theory have been presented by advancing research propositions: what effect does a market orientation have on business performance as measured by multiple performance indicators, and does the linkage between a market orientation and business performance depend on the external environment, such as competitive intensity, market volatility and technological turbulence?

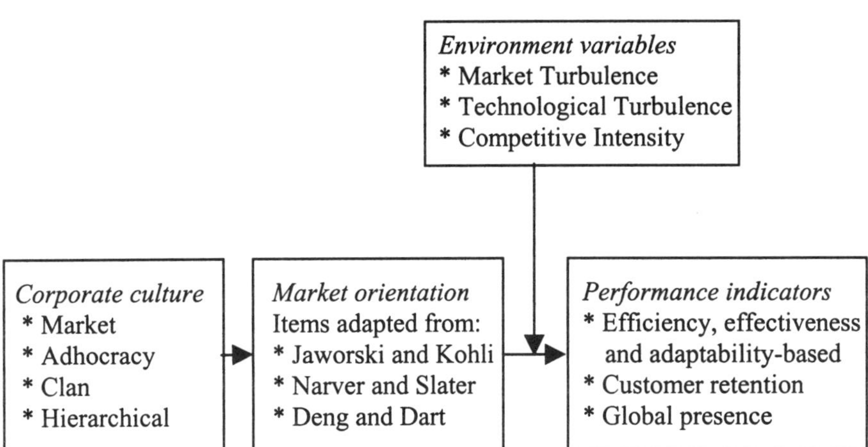

Figure 1.1 The theoretical framework

The conceptual foundation for the theory is offered by constructing an integral framework which includes the antecedents to market orientation and consequences of market orientation on business performance. Specifically, the following hypotheses are formulated for testing in the subsequent chapters.

H1: Market Orientation leads to a better Business Performance.

H2a: The lesser the extent of Market Turbulence, the greater the positive impact of Market Orientation on Business Performance.

H2b: The lesser the extent of Technological Turbulence, the greater the positive impact of Market Orientation on Business Performance.

H2c: The greater the extent of Competitive Intensity, the greater the positive impact of Market Orientation on Business Performance.

Further, a summary of the recent literature relating to market orientation and business performance is presented. The selection of the studies was primarily through their relevance which enlightened the concept of market orientation-business performance further in the setting of different industry or country. The criteria for selection were based on a judgment that the study would have an interest to marketing scholar interested in both market orientation-business performance relationship and the venue/environment of the study. No efforts were made to include all studies based on market orientation.

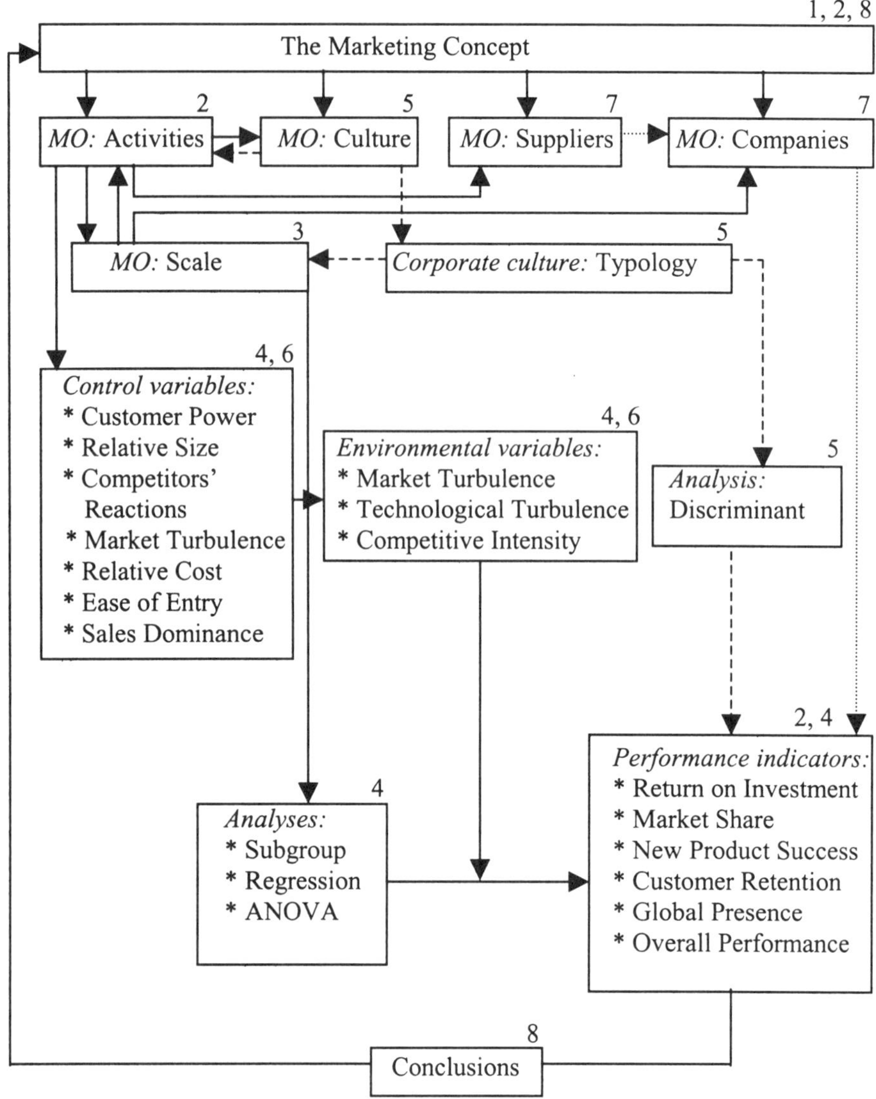

Source: The author

Figure 1.2 Organization of the book

Where, ——▶ Market orientation (MO) as activities, ----▶ Market orientation as a culture, and ⋯⋯▶ Market orientation case studies.

Chapter three complements the area of industrial marketing research where little work has been done to develop and test sector-specific market orientation scale. For the development of the market orientation scale, the paradigm suggested by Churchill (1979) is followed and described in detail. The comparative analysis among these existing market orientation scales (Narver and Slater, 1990; Jaworski and Kohli, 1993; Deng and Dart, 1994) has been performed for the generation of the scale items, as there is overlap of items on conceptual and operational basis. Further, keeping with the measurement models suggested by Bagozzi and Fornell (1982) and Bollen and Lennox (1991), attention was paid to the selection of items that were reflective in nature. Clearly, this was important because measurement model misspecification could undermine both construct validity and statistical conclusions. Data were collected from the British machine tool industry in the categories of SIC 3541 and SIC 3542 using single-informant postal survey methodology. Two-wave mailing yielded an effective response rate of 24 per cent. Preliminary results indicated that the 35-item market orientation scale is represented by a seven-factor solution, explaining 67 per cent of the variance in the scale. These factors were labelled as customer focus, competitor focus, responsiveness, customer satisfaction focus, market information, marketing focus and negligence focus. The reliability of each factor was measured by Cronbach Alpha and split-half Alpha methodology. Finally, some aspects of validity of the scale were examined by employing qualitative and quantitative methodologies to have greater confidence in the development of the scale and research findings.

Chapter four investigates empirically the relationship between market orientation and business performance. Business performance is measured by return on investment, sales growth, market share growth, customer retention rate and global market presence of companies. Further, the chapter examines whether the hypothesised relationship is moderated by external environment variables, such as market turbulence, technological turbulence, and competitive intensity. Subgroup analysis (SGA), moderated regression analysis (MRA) and analysis of variance (ANOVA) techniques have been employed to detect the effects from these environments on the market orientation-business performance relationship. Because multiplicative terms could introduce multicollinearity in moderated regression equation, all predictor variables, as suggested by Jaccard et al. (1990), were mean-centred before multiplicative interaction term was formed to reduce the effects of multicollinearity and standard error. Multicollinearity did not appear to pose any bias in the linear regression analysis. Next, a parsimonious model representing the market orientation scale has been developed. It was discovered that market orientation, in the context of the machine tool industry, is best represented by factors relating to customer, competitor, satisfaction, and marketing orientation only. Finally, the effect of each individual factor has been examined on overall performance of companies.

Chapter five focuses on the corporate culture as an antecedent to the market orientation, and explores the linkage between the dimensions of the corporate culture and the market orientation. The corporate culture scale was adapted from Deshpande et al. (1993) which was drawn upon the studies of Quinn (1988) and Cameron and Freeman (1991). Corporate culture has four dimensions of distinct culture: market culture, adhocracy culture, clan culture, and hierarchical culture. And each dimension has four items representing the distinct form of corporate culture. In the remainder of the chapter, the dimensions of corporate culture are conceptually related to the level of market orientation of companies, leading to the formulation of the following hypothesis:

> *H3:* The degree of market orientation of machine tool manufacturing firms operating in the UK varies respectively from highest to lowest according to the type of corporate culture as follows: Market, Adhocracy, Clan and Hierarchical.

Chapter six presents a general overview of the machine tool industry with the aim to familiarise the reader of the definition, function and evolution of machine tools including flexible manufacturing system. Next, the chapter reports and comments on the geographical location of British machine tool manufacturers and their concentration with employment characteristics. Pertinent economic figures such as balance of trade, share of world production and export of machine tools, and leading export and import markets in the world for machine tool manufacturers. The chapter concludes by shedding light on the current trend of the machine tool technology.

Chapter seven strengthens the design of the study by employing triangulation and case study methodology. The chapter explores the market orientation practices adopted by select machine tool companies and examines if the level of market orientation perceived by suppliers is the same as their respective customers. There are four companies in the study representing four distinct cases. The focus of the case study is at a corporate level. Each case represents assimilation of managerial thinking under a unique set of business conditions, such as market turbulence, technology turbulence, competitive intensity, competitors' concentration, among others. The emphasis of the cases is on the understanding the corporate executives' attitude towards their customers rather than on any functional department of the business. However the marketing stance was further explored by conducting a limited survey. A gap analysis was performed to detect differences between the perceived attitude of managers and their respective customers towards market orientation. Specifically, the following hypothesis was tested: The customers' perception of the market orientation is more important than suppliers' own perception of market orientation in explaining the suppliers' own business performance.

Chapter eight reports the results of study along with discussions of the findings. Finally the chapter concludes with acknowledging limitations, and highlighting managerial implications and indicating future research directions.

Chapter 2

Marketing Concept and Market Orientation

Introduction

The concept of market orientation is believed to be the cornerstone of marketing management because of the prevailing wisdom that having a market orientation is critical to superior performance and long-term success in today's highly competitive business environment. Despite the importance of market orientation in modern marketing management and strategy literature, systematic studies which aim to provide a rich understanding of the theory have only recently begun, following the seminal works of Kohli and Jaworski (1990) and Narver and Slater (1990).

The chapter begins with the introduction to the marketing concept and the development of market orientation theory. Theoretical and conceptual foundations for the theory have been presented by advancing research propositions. Specifically, the chapter addresses two conceptual questions: what effect does a market orientation have on business performance as measured by multiple performance indicators; and does the linkage between a market orientation and business performance depend on the external variables? Finally, a summary of pertinent literature relating to market orientation and performance is presented.

Marketing Concept and Market Orientation Theory

For more than three decades, the marketing concept has been a topic of interest to both academics and practitioners. In recent years, there has been a revival of interest in market orientation and its links with the business performance of companies. Marketing journals are full of articles, which advocate adoption of the marketing concept by firms.

The General Electric Company is usually acknowledged as the first firm to have systematically restructured its business operations in accordance with the marketing concept. In 1950, Ralph J. Carliner, an executive of the General Electric Company reengineered a master plan for the company's functional departments with the aim to develop and distribute products and services to satisfy consumers' needs. Because consumer needs were considered as the fundamental issue for all divisions that wished to implement the marketing concept, it was necessary for the divisions to consider what the consumers wanted. Certainly, these needs were important in designing and producing products (Business Week, 1950).

The reason why General Electric received much attention during this period was the fact that the main elements of the marketing concept, such as identification of current and future needs were understood and practiced intuitively. For the first time the marketing concept was well articulated. Further, the concept was employed at the early stages of the planning process of the product rather than at the end of the production cycle. Therefore, to satisfy the needs of customers in a better way, marketing decisions became a part of all major decisions. This integrated concept of customer-oriented marketing was viewed as the fundamental marketing concept guiding and coordinating the operation of the entire organization.

Keith's (1960) article on the marketing concept, one of the earliest and most popular, advocates the adoption of the marketing concept in an applied setting. According to the article, an organization evolves through three managerial stages: production, sales and market orientations. The implication is that this evolutionary process is correct for all organizations, because the goal of any organization, intending to be a viable entity in a competitive marketing dominated environment, is to generate profit (Houston, 1986).

Throughout marketing literature, the adoption of the marketing concept has been a foundation for successful business performance. Although the manifestations of adoption have not yet achieved clarity of definition (Kohli and Jaworski, 1990), different authors have defined the marketing concept in different ways (Kotler, 1980; McCarthy and Perrault, 1984; Houston, 1986). There appears to be no agreement as to the definition of the marketing concept; however, the underlying theme encompasses customer groups with similar demand characteristics and satisfying each of them with a specific marketing mix while achieving organizational objectives (Kotler, 1988). The main emphasis is on customers' present and future needs and their satisfaction, which is often a required ingredient to achieve competitive advantage (Konopa and Calabro, 1971; Larreche, 1985; Houston, 1986). There have been attempts to redefine the marketing concept. The marketing concept is a result of the more precise definition of marketing (Levitt, 1960; Kotler and Levy, 1969; Barksdale and Darden, 1971; Crosier, 1975; McCarthy and Perreault, 1984; Houston, 1986; Webster, 1988). For instance, Felton (1959) describes the marketing concept as a corporate state of mind that insists on the integration and co-ordination of all departmental functions for the basic objective of producing maximum long-range corporate profit. Further, it is defined as the orientation of the organization's attitude towards its customers (Felton, 1959; McNamara, 1972; Levitt, 1960). Along the same line of thought, Ferrell and Lucas's (1987) definition of marketing concept is that marketing is a process of planning and executing the concept of pricing and promotion, and distributing goods and services to create an exchange that satisfies individual and organizational objectives.

On the other hand, Konopa and Calabro (1971) define the marketing concept as the external consumer orientation as opposed to internal orientation around the production function; profit goals as an alternative to sales goals; and complete integration of organizational and operational efforts.

Although focus on customer, emphasis on profit and integration of departments in organizations have been stressed by many authors, it seems that the term marketing concept has become synonymous with the term 'customer orientation and customer satisfaction'. Certainly a satisfied customer is a constant source of profit for an organization. This notion is more forceful when several organizations offer similar products and services at very competitive prices. In which case, the customer orientation becomes the determinant of implementation of the marketing concept. Therefore, it is paramount that an organization that attempts to be oriented according to the marketing concept aims all its efforts at satisfying its customers at a profit (McCarthy and Perreault, 1984). Indeed, the marketing concept calls for most of the efforts to be spent on discovering the wants of a target audience and then creating the goods and services to satisfy them.

The marketing concept is essentially a business philosophy or a policy statement (Barksdale and Daren, 1971; McNarma, 1972). In keeping with the tradition and consistent with the definition given by aforementioned authors, McCarthy and Perreault (1984) and Kohli and Jaworski (1990) used the term market orientation to mean the implementation of the marketing concept. Hence, a market-oriented organization is the one whose actions are consistent with the marketing concept. However, it must be noted that a firm's ability to actually develop and implement a market-oriented philosophy may not be an easy task (Lichtenthal and Beik, 1984)

The Market Orientation

Implementation of the marketing concept is market orientation (McCarthy and Perreault, 1984). It is also stated that the marketing concept needs a better prescription for its implementation (Houston, 1986; Webster, 1988). For example, marketing implementation calls for the transition from an internally oriented R&D, finance, production, personnel or sales-focused company to a company that is oriented to act in accordance with the marketing concept. A growing number of business periodicals suggest that only a few companies have become fully market-oriented (Shapiro, 1988). An underlying theme in the literature survey is the need for each functional area to co-ordinate activities more effectively towards the common objective of satisfying the customer. Much of the literature analyzes the importance of the marketing department's interaction with other functional departments within the organization, such as R&D (Sounder, 1980; Shanklin and Ryans, 1984; Gupta et al., 1985-1986; Wilson and Ghingold, 1987; Lucas and Bush, 1988); manufacturing quality (Kohontek, 1988; Cravens et al., 1988; Atuahene-Gima, 1995); and sales (Kotler, 1977). This highlights the extent to which the marketing department is interdependent on other business functions. The loss of communication among different functions exists as a result of the differing nature of the functions, their tasks and responsibilities and how important they are perceived within an organization. Analysis of different functions in an organization often reveals the underlying nature of the conflicts leading to the loss of communication between these functions and the marketing department, thus increasing the difficulty in implementing the marketing concept. The following are the typical acrimonious statements of functional heads that defend their function or unit and put forward a solution from their particular point of view, which may result in lack of cohesion among functions. Some of the typical statements highlighted by Shapiro (1988) are as follows:

* Sales VP: We need more sales people. We are the ones who are close to the customers. We have to have more call capacity in the sales force so we can provide better service and get new product ideas into the company faster.
* Manufacturing VP: We all know that our customers want quality. We need more automated machinery so we can work to closer tolerance and give them better quality. Also, we ought to send our whole manufacturing team to Crosby's Quality College.
* Research and Development VP: Clearly we could do much better at both making and selling our products. But the fundamental problem is a lack of new products. They are the heart of our business. Our technology is getting older because we are not investing enough in R&D.
* Finance VP: The problem is not enough resources; there are too many resources misspent. We have got too much overhead. Our variable costs are unreasonable. And we spent too much on R&D. We do not need more; we need less.

All the above-mentioned statements are varied in nature without a clear market-oriented direction. This is a distinct impediment to the implementation of market orientation. Shapiro (1988) further argues for effective communication and co-ordination across departmental barriers to become market-oriented. Shapiro (1988) identified some key factors for companies to be market-oriented. They are:

* Having enough information on the buyers' characteristics and how they influence the role of each individual department.
* Making strategic decisions, taking all functional departments into account, to provide solutions for buyers.
* Having well co-ordinated divisions, making decisions and exercising them with a sense of commitment.

On the other hand, Masiello (1988) identified few factors that serve as barriers to a company attempting to become market-oriented:

* Most functional areas in the companies do not understand the concept of being truly driven by the market and customer needs. If a marketing plan does exist in the company, it is frequently not well communicated to those who must carry it out: salespeople, customer service representatives, clerks, order processors, production line workers, etc.
* Most employees do not know how to discharge their functional responsibilities into market and customer-responsive actions; conversely, they do not know how to look at the market from their functional perspective and do not recognize opportunities in the market.
* Most functional areas do not understand the role of other functions in the company. Hence, they do not know what information they should communicate to the other functions about the customer and what information they should seek from those other areas. Instead, they operate from functionally defensive positions, protecting the integrity of their own goals in terms of, for example, sales volume quotas, productions, defect ratios, etc.

* The employees in each functional area do not have meaningful input into the marketing decision of the company, even although they are often close to the operational characteristic of the marketplace. For example, how many of your customer service people:

  - Know about your customers' problem; and
  - Sit on the strategic or marketing planning committee?

The literature also stresses that chief executive officers (CEO) play an important role in the development of market-oriented consciousness within a company, as they are the ones who understand the situation of the marketplace from a bird's eye point of view. Canning (1988) focuses on identifying the role of a CEO in an organization and offers several questions to test whether a company is market-oriented in terms of marketing consciousness.

- CEO's ability to reflect an understanding of the reality of the marketplace.
- CEO's role of corporate strategy information system.
- CEO's role of new product development.
- CEO's endeavour of corporate culture in establishing a firm's market orientation.
- CEO's role of the sales staff.

The essence is that every employee needs to look for one of the marketing elements in their role. Based on the discussion above, it is not easy to clearly define market orientation, yet it seems as though most of the authors agree on the need to be active to the customers' needs. Therefore, market orientation demands a market consciousness that goes beyond normal functions and encompasses the whole organization. This means a clear understanding of customer needs, good leadership, an appropriate culture and a clear awareness of the external environment.

Market Orientation and the Definition

Thinking in terms of the overall market and not marketing *per se* is an essential ingredient in the recipe for the creation of competitive advantage. Although, competitive advantage can be created based on organizational structure, market power, economies of scale, product range etc., recently there has been a significant shift in the ability of firms to create and consistently deliver a superior value to its customers by becoming market-oriented. The need for all types of organizations engaged in economic activities to adopt market orientation (versus product or sales orientations) is now a well-settled issue. A business is said to be market driven when its culture is systematically and entirely committed to continuous creation of superior customer value. Kohli and Jaworski (1990) conclude that market orientation provides a unifying focus for the efforts and projects of individuals, thereby leading to superior performance. Hence, the survival and eventual growth of any organization hinges on the adoption of a market orientation philosophy. Often organizations are unclear as to what market orientation is and what it means. The gap between the idea and practice tends to be greater in the case of large organizations because a number of administrative layers create a distinct impediment to the flow of information between customers' needs and the decision makers who run these organizations.

For an organization to have a strong market orientation, it must direct its resources in a particular direction with the aim to monitor continuously the changing expectations of their customers. Organizations must look out for any possible opportunities that can change its future decisively. For example, to create a superior value, it is essential to understand not only a customer's present needs but also potential needs. It is also important that sellers understand a buyer's entire value chain. At any point in the supply chain, the value for the buyer can be created by two ways: either by making a more effective presence in the target market or by creating more efficiency in its operations. Certainly, employees of market-oriented businesses have to spend time with their customers (Slater and Narver, 1994).

Although the subject of market orientation has been the focal point of the marketing theory for many decades (Levitt, 1960; Kotler, 1977; Shapiro, 1988; Webster, 1988), recently researchers have attempted to construct a theory on the antecedents and consequences of market orientation. The origins of market orientation can be traced back to the philosophy of the marketing concept, which holds that the ultimate goal of an organization is to fulfil customer needs for the purpose of maximizing profits. Drucker (1954) asserts that creating a satisfied customer is the only justifiable definition of a business. Kohli et al. (1993) have developed a scale to measure the market orientation and tested its effect on business performance.

While the marketing discipline was attracting increasing attention among scholars, a realization of the significance of the customer was also emerging within the marketplace. McKitterck (1957) of General Electric augmented the marketing concept by advocating that the purpose of a business should be to respond to the customer rather than to attempt to change the customer to suit the firm's purposes. During the same era, Levitt (1960) contended that the business definition should be on customer needs instead of a specific offering employed to meet those needs. Perhaps, the notion of market orientation came into force following the radical changes in both domestic and global economies in the 1970s. There was a resurgence of interest in the marketing discipline as businesses sought to develop and maintain a focus on the customer. This pursuit was accentuated by writers such as Peters and Waterman (1982) who identified closeness to the customer as the chief common characteristics of the best US organizations, while in the UK, Hooley and Lynch (1985) established marketing excellence as the distinguishing feature of the best performing firms. Hence, it is not surprising that the significance of customer-based information became a key feature of an appropriate conceptualization of market orientation in the late 1980s and early 1990s.

Recent development of marketing theory has shown that there are three main conceptualizations of market orientation (Narver and Slater, 1990; Jaworski and Kohli, 1993; Deng and Dart 1994). Kohli and Jaworski (1990) have defined market orientation as three components: generation of market intelligence, dissemination of the intelligence and organization-wide responsiveness to it. Market intelligence pertains to current and future customer needs and preferences while taking into account how these needs and preferences may be affected by exogenous factors, such as government regulation, technology, competition, etc. Intelligence dissemination refers to the extent to which the intelligence is disseminated throughout the organization to the relevant individuals and departments. Finally, responsiveness is the action taken in response to intelligence that is generated and disseminated. Responsiveness to market intelligence has two dimensions: response design, i.e. development of a plan; and response implementation, i.e. execution of a plan. Figure 2.1 shows the conceptual framework of Kohli and Jaworski (1990).

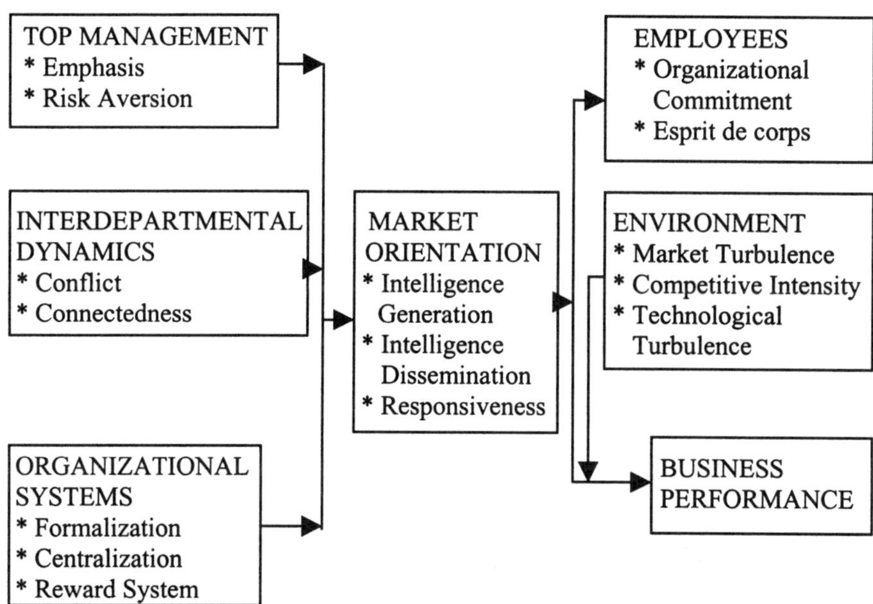

Figure 2.1 Conceptual framework of Kohli and Jaworski

Source: Kohli and Jaworski (1990)

On the other hand, Narver and Slater (1990) define market orientation as three behavioural components consisting of customer orientation, competitor orientation and interfunctional co-ordination. Customer and competitor orientations include all the activities relating to the information generation about the buyers and competitors in the designated target area, whereas interfunctional co-ordination is based on the utilization of company resources effectively in creating a superior value for the target customers. Because any individual in any function in a seller firm can potentially contribute to the creation of value for buyers (Porter, 1985), any company can be on the path to becoming market-oriented. Figure 2.2 shows the conceptual framework of Narver and Slater (1990).

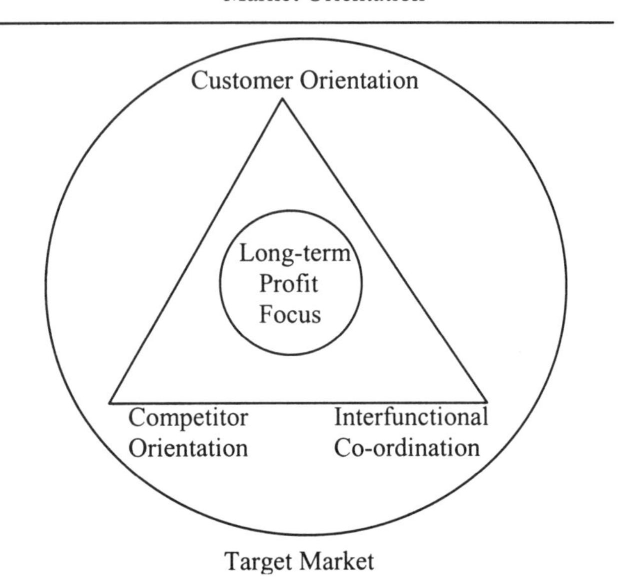

Figure 2.2 Conceptual framework of Narver and Slater

Source: Narver and Slater (1990)

Ruekert (1992) defines market orientation as the degree to which an organization obtains and uses information from customers, develops a strategy and implements that strategy by being responsive to customer needs and wants. Deshpande et al. (1993) describe customer orientation as a set of beliefs that puts the customer's interest first while not excluding those of all other stakeholders, such as owners, managers, employees etc. in order to develop a long-term profitable enterprise

Deng and Dart (1994) have developed a scale to measure market orientation, which is very similar to that of proposed by Narver and Slater (1990). Deng and Dart's (1994) scale includes an additional dimension to the scale - profit emphasis, considering it to be the main objective of a firm as profit orientation is an inherent practice in the day-to-day operation of most successful businesses. This perspective reflects a divergence in perspectives. Figure 2.3 shows the conceptual framework of Deng and Dart (1994).

Further, scholars have ventured to capture the domain of market orientation from different perspectives, e.g. organizational strategy (Ruekert, 1992), international marketing (Cadogan and Diamantopoulos, 1995), manufacturing sector (Singh and Ranchhod, 1998), retail industry (Soehadi et al. 2001), and others.

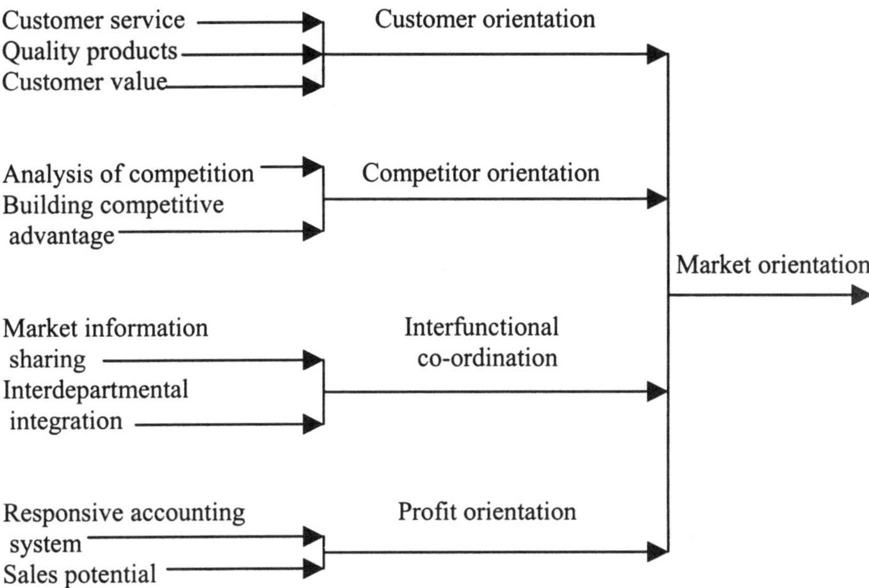

Figure 2.3 Conceptual framework of Deng and Dart

Source: Deng and Dart (1994)

There have been challenges to some of the common assumptions underlying development of a market orientation scale (Shapiro, 1988). While the need to pay attention to serve the needs of customers is not disputed, some researchers have suggested potential serious shortcomings in the present conceptualizations of market orientation (Chang and Chen, 1994; Siguaw et al., 1994; Siguaw and Diamontopoulos, 1995). For example, Meziou (1991) points out that these scales have frequently been insufficient to capture the comprehensive nature of a truly market-oriented philosophy. Some authors have also argued that adoption of a market-oriented business philosophy is only essential for survival in a very competitive environment (Levitt, 1960; Kotler, 1977; Kotler and Andreasen, 1987), whereas in a stable environment the concept may have lesser importance.

Such challenges have prompted researchers to consider the methodological issues (McKitterick, 1957; Churchill, 1979; Houston, 1986; Lusch and Laczniak, 1987) associated with the measurement of the market orientation scale. More recently, there has been concern with market orientation's alleged overemphasis on customer orientation (Narver and Slater, 1990). It is argued that becoming excessively customer-oriented may lead an organization to make only incremental improvements to existing product lines; therefore, this kind of approach may be detrimental to innovativeness. The fact that most technological advances, such as in the case of the technologically driven machine tool industry, are well beyond the realm of ordinary customer perception (Bennett and Cooper, 1981), a company could be led to applying only short-term strategies which may result in the production of *me-too* products. This strategy will also call for a massive advertising budget for the promotion of the *me-too* products. Therefore, in light of the above arguments, market orientation needs to be redefined whereby its focus should allow for a better balance between market considerations (external) and organizational capabilities (internal) to meet those needs.

Although these three market orientation scales are different in terms of the selection of items representing these scales, there is clearly an overlap on a conceptual and operational basis. Cadogan and Diamantopoulos (1995) have performed a comparative analysis between the components of Kohli and Jaworski and the components of Narver and Slater (1990) and have shown the conceptual and operational overlap between these two scales.

The domain specification of market orientation seems to be complex, as there is no single definition of the philosophy of market orientation. The review of literature reveals that there are a number of meanings ascribed to market orientation. For example, Konopa and Calabro (1971) place greater emphasis on customer than production and cost related activities whereas according to Felton (1959) and McNamara (1972), the involvement of marketing executives in the strategic decision making process and the integration of activities within the marketing function is a central issue for a company to be market oriented. Although, these aforementioned authors differ in their preferred conceptualizations, it is evident that there are three main themes underlying market orientation: customer focus, i.e. information generation pertaining to customers; competitor focus; and responsiveness, i.e. dissemination of information pertaining to customers across the functional departments with the view to meet customer needs quickly by having effective interfunctional co-ordination of the concerned departments.

In the study, following the interviews with the marketing directors of machine tool companies and literature search, it was found that emphasis on customer satisfaction has increased over the last decades. Measuring customer satisfaction in connection with customer loyalty, i.e. willingness to purchase from the same company, is an important tool for improving the competitiveness of companies. This reflects the need for retaining customers and developing long-term relationships. For some companies, particularly SMEs, loyal customers are paramount for existence (Davis et al., 1993).

Retaining customers can have a significant positive impact on profitability of companies (Singh, 1998). Studies have shown that retaining an additional two to five per cent customers can improve profits significantly in the same manner as cutting costs by ten per cent (Reichheld 1990; Power et al. 1992). Most researchers (Oliver et al., 1989; Wilkie et al., 1980, Singh and Ranchhod, 2004) and practitioners agree that satisfaction occurs when purchase expectations are met, i.e. attributes associated with products are the ones desired by the customers. This implies that companies should, in addition to being customer- and competitor-oriented, be satisfaction-oriented as well to meet purchase expectations. Dissatisfaction is the result of unconfirmed expectations. Marketers who understand the impact of customer satisfaction on business performance will want to secure future sales orders on the basis of the recommendations of currently satisfied customers of the products. In this respect, current buying decisions often affect future purchase decisions.

Therefore, in the context of the machine tool industry, the author defines market orientation as 'the set of activities co-ordinated in such a way that derives customer satisfaction through superior performance of products (machines) and related services (training, maintenance, etc.) while still being competitive (price, responsiveness, delivery, etc.) in the marketplace'.

For the purpose of this study, the author has taken the view of the combined components of the market orientation definitions given by the three (Narver and Slater, 1990; Jaworski and Kohli, 1993; Deng and Dart, 1994) to specify a new domain of market orientation except profit emphasis. The author believes that profit is the outcome of the adoption of the market orientation concept; therefore, should be treated as a behavioural component of market orientation (Levitt; 1960 and Narver and Slater, 1990).

Theoretical and Conceptual Foundations

Market Orientation and the Machine Tool Industry

The key to growth of the UK's manufacturing industry, especially the small- and medium-sized enterprises (SMEs), which are the backbone of the economy, is the investment in the latest technology. With the arrival of a single market in Europe and the rapid development of the Far Eastern economies, UK manufacturers need to improve their productivity and competitiveness by investing in technology. In the USA, Japan and Germany, consistently high levels of economic growth have been achieved over the long-term with high levels of investment in the technology. It is no coincidence that the strongest manufacturing nations also have the strongest machine tool industries.

The machine tool industry in the UK is a mature industry with many well-established companies, e.g. Bridgeport, 600 Group, and others. In recent years, the high rate of technological advancement has raised market entry barriers, which is marked by high levels of investment in both plant and machinery. Ashburn (1993) has shown that investment in R&D is essential for long-term success. Reduced profit margin has caused many well-known companies to face increasing financial difficulties and in some cases even shut downs, e.g. Shand Ltd., Buchler Phillips, Southern Tools Ltd, and others. The machine tool industry is characterized by a large number of SMEs, manufacturing a wide variety of types and sizes of products. Several small firms are subsidiaries of large firms. These small firms either manufacture the accessories, such as electric drives and CNC control devices for the large machine tool manufactures, or they supply to them after having them imported/assembled at low cost in Korea, China or India. These SMEs have simple organizational structures and a limited range of products and customers. These attributes may enhance the ability of smaller firms to create a market-oriented culture without resorting to formal mechanisms (Pelham and Wilson, 1996). More discussion about the British machine tool industry is presented in Chapter six.

There is evidence that internal organizational factors and external market forces may exert a considerable influence on the performance of SMEs. For example, a lack of planning, and under capitalization have frequently been advanced in literature as the most critical determinants of small firm failure (Robinson and Pearce, 1984). Given several SMEs are noted for their lack of long-range focus (Gilmore, 1971), strategic orientation and systematic decision-making (Robinson, 1982; Sexton and Van Auken, 1982), it may be argued that market orientation could be a critical determinant of their success. Further, such firms generally lack the financial resources to explore other sources of business profitability, such as research and development, competitive advantage, low cost leadership or skilled staff to develop effective planning strategies (Pelham and Wilson, 1996). An indication of this possibility comes from Narver and Slater's (1990) results, which indicate that larger Strategic Business Units (SBU) with low market orientation, but low cost advantages, outperform smaller SBUs with medium levels of market orientation, but not smaller SBUs with a high level of market orientation. A market-oriented strategy could provide firms with organization-wide focus for formulating objectives, making quality decisions, guiding actions and directing resources.

Market Orientation and Business Performance Relationship

Market orientation should enable companies to pursue consistency (Rumelt, 1981) in their actions to fulfil customers' needs. If actions are coherently guided by values aimed at satisfying customers, there should be consistency in the decision-making process across functions throughout the firm leading to consistency in product development. Firms that endeavour to understand customer needs with the view to satisfy those needs should develop products or services with relatively less defects which in turn should result in reduced costs and improved profitability. Further, the strategy of consistency is expected to reduce costs associated with rectifying errors that arise due to poor functional co-ordination; thus, reducing expenditure on human resources. Such efforts can facilitate implementation of market orientation, leading to higher performance of companies. Table 2.1 reports the results of some studies based on market orientation and its relationship with performance.

Table 2.1 Key findings between market orientation and performance link

Author	Country	Results	Performance Indicators
Narver et al. (1990)	USA	+	ROA
Ruekert (1992)	USA	+	Profitability
Diamantopoulos (1993)	UK	weak	Profitability, SG
J&K (1993)	USA	+	EDC, OC, OP
Orelowitz (1993)	Africa	ns	Profitability
Chang et al. (1994)	Taiwan	+	Profitability, Quality
Slater et al. (1994)	USA	+	NPS, ROA, SG
Atuahene-Gima (1995)	Australia	+	NPS
Caruana (1995)	Malta	+	ROCE, OP, SG
Greenley (1995)	UK	None	NPS, ROI, SG

Bhuian (1996)	Saudi	+	Organizational performance
Langerak (1996)	Holland	ns	Competitive advantage
Pelham et al. (1996)	USA	+	NPS, Profitability, Product quality
Pitt et al. (1996)	UK	+	OP, ROCE, SG
Appiah-Adu (1997)	UK	+	NPS, ROI, SG
Appiah-Adu (1998)	Ghana	ns	SG, ROI
Appiah-Adu et al. (1998)	UK	mixed	SG, ROI
Singh (1998)	UK	+	SG, NPS, Market Share, ROI
		ns	Overall performance
Craven et al. (2000)	USA	+	Brand valuation
Loubser (2000)	S. Africa	+	ROE, Total Assets
Sin et al. (2000)	China	+	SG, OP, Customer Retention
Grewal et al. (2001)	Thailand	-	After Crisis
Soehadi et al. (2001)	Indonesia	+	Retail Performance
		+	Supplier partnership
Vazques et al. (2001)	Spain	ns	Innovation rate
		+	Firm innovativeness
		+	New Product Innovativeness
		ns	Performance
Gainer et al. (2002)	Canada	+	Customer Satisfaction
		+	Peer Reputation
		+	Growth in Resources
Matsuno (2002)	USA	+	Market Share
		+	ROI
		+	% of new product sales to total sales
Tay et al. (2002)	UK	+	Performance
Delbaere et al. (2003)	Canada	ns	Revenue
Singh (2003)	India	+	Return on investment
		+	Customer Retention
		ns	Global Presence

Source: The author

Key: ROA=Return on Assets, EDC=Esprit de corps, OP=Overall Performance, OC=Organizational Commitment, ROE=Return on Equity, SG=Sales Growth, NPS=New Product Success, ROCE=Return on Capital Employed, and ns=Not Significant

While the impact of market orientation upon performance may be indirect, for example, via new product development (Cooper, 1984), sales growth, market share (Buzzell et al., 1975; Venkatraman and Prescott, 1990), relative product quality (Phillips et al., 1983; Buzzell and Gale, 1987), Innovation characteristics (Atuahene-Gima, 1996), Entrepreneurial proclivity (Matsuno et al. 2002), knowledge management (Delbaere, 2003), a direct positive impact of market orientation on business performance is also detected (Narver and Slater, 1990; Ruekert, 1992; Kohli and Jaworski, 1993; Chang and Chen, 1994; Slater and Narver, 1994; Caruana, 1995;

Greenley, 1995; Pitt et al., 1996; Pelham 1996; Appiah-Adu, 1997; Appiah-Adu and Singh, 1998; Loubser, 2000, Sin et al., 2000; Vazques et al., 2001; Gainer and Padanyi, 2002, Singh, 2003).

Pelham and Wilson (1996) provide empirical evidence for the association between market orientation and performance in small firms in the US, which mirrors the support provided by Slater and Narver (1994) in the SBUs of large organizations. Their results indicate that market orientation has a significant impact upon various dimensions of business performance.

Based on interviews with the managers of the machine tool industry, it was evident that market orientation may not have a significant effect on profitability in the short-term, but it may have an impact on profitability in the long-term. The argument to support this notion is that companies may not necessarily make significant profit by selling machines only, but can generate subsequent profits by providing customers with services after sales (SAS). This SAS orientation can have a significant impact on customer satisfaction, which may lead to higher business performance of companies.

This study tests the impact of market orientation on business performance in the machine tool sector in response to the call for the need to test the relationship in a high-tech industry (Slater and Narver, 1994). Hence, the central hypothesis to be tested is:

H1: Market Orientation leads to better Business Performance.

Performance Measurement Indicators

Performance measurement is an important issue in the field of management science. Although, there is a debate as to the reliability and validity of performance indicators, there is no substitute for their role in managerial decision-making process (Hofer, 1983). Such indicators provide managers with better insights into planning and control of organizational performance. The debate on the selection of performance measurement indicators is inconclusive. The most commonly used performance measurement indicators are: return on investment (ROI), net profit, liquidity, leverage ratio, gross and net contribution margins, market share, sales growth, turnover of personnel, among others.

Multiple measures of performance are frequently used in marketing research. Walker and Ruekert (1987) suggested three specific attributes for the measurement of performance: effectiveness, efficiency and adaptability. In this study, ROI was used to measure the efficiency of the organization; sales growth and market share were used to assess the firms effectiveness and ability to achieve market power. Market share has a positive relationship with profitability (Buzzell et al., 1975; Szymanski et al., 1993). New product success (NPS) rate was used to measure adaptability. Two additional performance indicators were used: *customer retention* to assess the level of satisfaction and *global presence* to measure the level of export orientation. Table 2.2 shows the relationship between performance indicators and their attributes.

Table 2.2 Performance indicator attributes

	Efficiency	Adaptability	Effectiveness
ROI	*		
Sales Growth			*
New Product Development		*	
Market Share			*
Customer Retention		*	
Global Presence		*	

Source: The author

Recently, the *market share* as a proxy for profitability of companies has come under question (Jacobson, 1990) because increasing market share by spending more on advertising and promotion may come at the expense of short-term profit. On the other hand, finance-based measures, such as ROI may assess the efficiency of an organization in the short-term, but may undermine the development of a long-term strategy (Eccles, 1991). It may not be adequate to predict corporate excellence only (Chakravarty, 1986). Therefore, it is preferable to devise a multiple measure of performance and interpret the results based on one performance indicator in conjunction with others. Multiple measures are consistent with the theoretical viewpoint in that performance should be viewed as being multidimensional as it covers diverse purposes and types of organizations (Levin and Minton, 1986). Jaworski and Kohli (1993) found a relationship between self-reported measures of performance and market orientation, but not market share. The relationship between market orientation and market share was not supported. In this regard, it is important to note that business performance is a multidimensional construct and may be characterized in a number of ways, including effectiveness, efficiency and adaptability. Further, performance on one dimension may run counter to performance on another dimension.

Therefore, in the study, different measures were used to obtain a comprehensive view of the performance of the business while reducing the impact of individual bias of any particular dimension (Schlegelmilch and Ross, 1987; Shoham and Ross, 1993). While using multiple measures is theoretically sound, it may lack convergence (Bhargava et al., 1994). The lack of convergence has been addressed in the study by two methods: First, a choice was made among the competing measures to select the best measures. Some direct measures were used in validating the indirect measures (Jaworski and Kohli, 1993). Second, a factor analysis was employed to check the dimensionality of the performance indicators. Reliability and validity checks were performed on these measures.

Efficiency-Based Performance Indicator: Return on Investment

It could be argued that the main objective of any business is to be profitable (Felton, 1959; McNarma, 1972; Narver and Slater, 1990; Ruekert, 1992; Jaworski and Kohli,

1993; Chang and Chen, 1994; Slater and Narver, 1994; Pitt et al., 1994; Greenley, 1995; Pelham and Wilson, 1996). Kohli and Jaworski (1990) viewed profitability as a consequence of market orientation. In a similar study conducted by Slater and Narver (1993), the profitability was perceived as an objective of a business. Consistent with the views of Kohli and Jaworski (1990) and Narver and Slater (1990), it was proposed to test the following hypothesis taking ROI as one of the efficiency-based performance indicators:

> *H1a:* Market Orientation is positively related to Return on Investment (ROI).

Effectiveness-Based Performance Indicator: Sales Growth and Market Share

Considering, *sales growth* and *market share* as one of the effectiveness-based performance indicators, the following hypotheses were proposed (Slater and Narver, 1994; Greenley, 1995; Caruana, 1995; Pitt et al., 1996; Pelham and Wilson, 1996):

> *H1b:* Market Orientation is positively related to sales Growth.

> *H1c:* Market Orientation is positively related to Market Share.

Adaptability-Based Performance Indicator: New Product Success

Recent studies have shown that there is a positive correlation between *market orientation* and *new product success* (Slater and Narver, 1994; Greenley, 1995; Pitt et al., 1996). Some scholars argue that strict adherence to the principles of market orientation leads to poor innovation activities (Kirby, 1972; Tauber, 1974; Hayes, 1980; Bennet and Cooper, 1981), thus the poor performance of companies in the long-term. Others provide strong evidence for the potential benefits of market orientation to new product development (Bentley, 1990; Kohli and Jaworski, 1990). Parasuraman (1980) provides inconclusive evidence on the influence of market orientation on new product development activities.

In the context of the machine tool industry, NPS is particularly important to the success of key industries, such as aerospace and automotive sectors, which account for major machine tool purchases in the UK. For the development of new products, machine tool suppliers invest significantly in research and advanced machine tooling manufacturing technology to improve their productivity and competitiveness. The improved productivity and competitiveness is believed to have a positive impact on the performance of companies. Therefore, the hypothesis to be tested is:

> *H1d:* Market Orientation is positively related to New Product Success

Customer-Based Performance Indicator: Customer Retention

Return on investment and sales growth could be a result of customer retention. Customer retention is the direct consequence of customer satisfaction, which may be influenced by the quality of the product (Atuahene-Gima, 1996) and the quality of service provider (Innis and Lalonde, 1994), among others. The machine tool industry is considered to be a capital-intensive industry where the frequency of capital purchases is low. Often a machine has a life span of two to ten years depending upon the complexities and the degree of obsolescence of the technology. Given the long life cycle of machines, it is paramount that the end-users of machines are satisfied with the performance of the machines and with the quality of the service provided by the suppliers or the manufacturers of the machines, which means manufacturing machines that meet the exact technical specifications required by the end-users. This helps suppliers get repeat orders. The key criterion to get repeat order is to keep the customer satisfied, which can be accomplished by improving continuously the capabilities of existing machines, thus providing customers with superior quality products and services. Satisfaction has been found to result in customer retention, which involves purchasing from the same supplier, increased relationship and word-of-mouth recommendations. Therefore, customer retention may have a direct impact on the profitability of companies, as it reduces customer acquisition costs, lowers costs of serving repeat orders and improves mutual trust between buyers and sellers. Reichheld and Sasser (1990) found that five per cent growth in customer retention could enhance profit levels from 25 per cent to 85 per cent.

Customer satisfaction results in loyal customers who purchase repeatedly, contributing to profitability and growth (Hallowell, 1996). Therefore, taking customer retention as one of the customer-based performance indicators to assess the impact of market orientation, the following hypothesis was proposed:

H1e: Market Orientation is positively related to Customer Retention.

Export-Based Performance Indicator: Global Presence

Because very few countries build every type of machine tool, international trade is important. This is the key to the future development of industrialized economies, such as Taiwan, South Korea, China and India. Therefore, it is intended to test whether managers pursue market-orientated activities abroad in the same way as they do in their domestic markets (Cadogan and Diamantopoulos, 1995). Previous findings suggest that the managers lacked overseas market research activities (Hooley et al., 1983). This myopic approach towards the internationalisation of market orientation has resulted in numerous business failures (Hooley et al., 1983; Mayo, 1991; Taylor, 1992). Machine tool industry is characterized as an export-oriented industry as it often meets more than domestic requirements, which means that if a company can create a superior value for its foreign customers, it should be able to have competitive advantage over manufacturers for the supply of similar machines. As a result, a firm would have foreign presence through export orders. Further, as requested by foreign customers or dealers, several manufacturers are beginning to incorporate additional features to the machines to qualify these foreign markets.

Therefore, it was proposed to use the *global presence* as one of the export-based performance indicators to measure the impact of market orientation. The hypothesis to be tested is:

H1f: Market Orientation is positively related to Global Presence.

Moderators: the Exogenous Variables

Studies from several countries have examined the link between market orientation and business performance (Narver and Slater, 1990; Ruekert, 1992; Jaworski and Kohli, 1993; Deshpande et al., 1993; Chang and Chan, 1994; Slater and Narver, 1994; Atuahene-Gima, 1995, Caruana, 1995; Bhuain, 1996; Langerak et al., Netherlands, 1996; Pitt et al., 1996; Appiah-Adu, 1998; Appiah-Adu and Singh, 1998; Loubser, 2000; Grewal and Tansuhaj, 2001; Gainer and Panlette, 2002; Noble and et al. 2002; Tay and Morgan, 2002; Marjorie et al., 2003; Singh, 2003). Few studies have reported the impact of environmental moderators on market orientation business performance relationship (Jaworski and Kohli, 1993; Slater and Narver, 1994; Diamantopoulos and Hart, 1993; Atuahene-Gima, 1995; Greenley, 1995; Langerak et al., 1996; Appiah-Adu, 1997; Appiah-Adu and Singh, 1998; Matsuno and Mentzer, 2000; Grewal and Tansuhaj, 2001; Tay and Morgan, 2002; Marjorie et al. 2003; Singh, 2003).

Although market orientation is related to business performance in general, under certain conditions it may not be critical Kohli and Jaworski (1990). Managers of business operating under these conditions should pay close attention to the cost benefit ratio of a market orientation. Given the limited amount of resources available to increase market orientation, managers should carefully assess the expected benefits and cost of increasing the market orientation within the environment they operate. Managers should identify the external environmental factors under the influence of which the expected cost of increasing or maintaining a market orientation could not exceed its expected benefits. If the cost of increasing or maintaining market orientation exceeds expected profit, it indicates that the extent to which market orientation should be increased would depend upon the turbulence in the environment.

The level of environmental turbulence is critical to survival and growth of companies. It is particularly important in the case of SMEs because these companies are presumed to be sensitive to changes in business conditions. Despite the fact that SMEs have the ability to be flexible in terms of organizational structure and that they are more responsive to environmental changes, they are also challenged by unfavourable and hostile environments (Covin and Slevin, 1989). While it is demonstrated that market orientation improves business performance, it is also acknowledged that environmental uncertainties might affect the postulated relationship (Day and Wensley, 1988; Kohli and Jaworski, 1990). The potential effects of environmental influences are consistent with a long tradition of support for the notion that the environment moderates the strength of the market orientation business performance relationship.

External environments, such as the strength of the economy, market growth, market turbulence, competitive intensity, etc. have been established as factors which have impacts on business performance (Cooper, 1979; Miller and Freisen, 1983; Peterson, 1985; Miller and Toulouse, 1986). For instance, high competitive intensity may threaten companies, particularly SMEs, because of their limited resources to cope with the changes in the business environment. Therefore, the ability of companies to compete effectively in a hostile marketplace may become an important issue for the survival of companies.

Kohli and Jaworski (1990) found that business environments altered the relationship between market orientation and business performance. This finding supports the notion that firms will modify their degree of market orientation as changes take place in the marketplace. For example, the authors found that several managers believed that they could get by with a limited level of market orientation in markets where demand outstripped supply. This implies that market orientation has a relatively stronger impact upon performance under conditions of low demand. In other words, under certain conditions market orientation is not critical. In such environments, it is important for managers to evaluate the anticipated costs and benefits of improving their firms' market orientation. Subsequently, several recent studies have examined the influences of moderating variables upon performance. Jaworski and Kohli (1993) searched for moderating effects from *market turbulence*, *competitive intensity* and *technological turbulence* and found no evidence of these environments affecting the relationship between market orientation and business perormance. Diamantopoulos and Hart (1993) found competitive intensity to be an important environmental variable. Slater and Narver (1994) reported partial support for market growth, market turbulence and technological turbulence. Greenley (1995) identified effects from *market turbulence* and *technological turbulence*. Pelham and Wilson (1996) found that the degree of market orientation among small firms was not influenced by *market dynamism* (combined effects of market and technological turbulence) or competitive intensity. Further, Appiah-Adu and Singh (1998) found moderating effects from *market growth* on *sales growth* but no effect was found from *market dynamism* on *return on investment*. In recent studies, Grewal and Tansuhaj (2001) examined the moderating effects from competitive intensity, product demandand technological uncertainties on market orientation-after crisis link. Delbaere et al. (2003) tested the effect of knowledge management on market orientation-customer loyalty relationship. In transition economies, Singh (2003) found that the link between market orientation and business performance was stronger when competitive intensity was high and market dynamism was low. The moderator effects of environmental variables on market orientation business performance link are reported in Table 2.3.

Based on the long tradition of support for the theory of external moderating factors and the conflicting evidence of previous studies (Jaworski and Kohli, 1993; Diamontopoulos and Hart, 1993; Narver and Slater, 1994; Greenley, 1995; Atuahene-Gima, 1995; Pelham and Wilson, 1996; Appiah-Adu and Singh, 1998, Grewal and Tansuhaj, 2001; Tay and Morgan, 2002; Delbaere et al. 2003; Singh, 2003), the search for possible moderators was intended in the study. For the purpose, three external variables: market turbulence, technological turbulence and competitive intensity have been utilized to examine the possible effects of these external variables on the market orientation-performance relationship. Market growth was not included in the study because it is supposed to be related to the strength of economy. The Intention of the study was to include only those variables that were directly related to marketing.

Table 2.3 Results of moderators on market orientation performance link

Author(s)	Moderating Effect	Performance Measures
Business Strategy		
Matsuno and Mentzer (2000)	Yes	Business Performance
Competitive Intensity		
Jaworski and Kohli (1993)	No	Performance
Diamantopoulos and Hart (1993)	Yes	Sales Growth
Slater and Narver (1994)	No	Performance
Atuahene-Gima (1995)	Yes	New product success
Bhuian (1996)	Yes	Organizational Performance
Langerak et al. (1996)	Yes	Competitive Advantage
Appiah-Adu (1997)	Yes	New Product Success
Appiah-Adu (1998)	Yes	Sales Growth
Singh (1998)	Yes	Sales Growth
Grewal and Tansuhaj (2001)	Yes	After Crisis
Tay and Morgan (2002)	No	Business Performance
Singh (2003)	Yes	Return on Investment
Knowledge Management		
Delbaere et al. (2003)	No	Customer Loyalty
Market Dynamism		
Singh (2003)	Yes	Customer Retention
Market Growth		
Diamantopoulos and Hart (1993)	Yes	Sales Growth, Profit
Slate and Narver (1994)	Yes	Sales Growth
Appiah-Adu (1997)	Yes	Sales Growth
Appiah-Adu and Singh (1998)	Yes	Sales Growth
Market Turbulence		
Diamantopoulos and Hart (1993)	Mixed	Performance
Kohli and Jaworski (1993)	No	Performance
Slater and Narver (1994)	Yes	Return on Assts
Greenley (1995)	Yes	Return on Investment
Appiah-Adu (1997)	Yes	Return on Investment
Appiah-Adu (1998)	Yes	Return on Investment
Singh (1998)	Yes	Return on Investment
Appiah-Adu and Singh (1998)	No	Return on Investment
Tay and Morgan (2002)	No	Business Performance
Product Demand		
Grewal and Tansuhaj (2001)	Yes	After Crisis

Technology Change		
Kohli and Jaworski (1993)	No	-
Slater and Narver (1994)	Yes	New Product Success
Greenley (1995)	Yes	New Product Success
Bhuian (1996)	Yes	Organizational Performance
Appiah-Adu (1997)	No	Performance
Singh (1998)	Yes	New Product Success
Grewal and Tansuhaj (2001)	Yes	After Crisis

Source: The author

Market Turbulence as a Moderator

Market turbulence has been defined as the extent to which customer needs have changed and the extent to which marketing operations have changed as a consequence (Miller, 1987). It is also described as the speed of change in the marketing tactics required to cater for customer needs. Dess and Beard (1984) describe both environmental turbulence and dynamism as changes that are unpredictable and difficult to plan.

It is expected that market turbulence would influence the level of a firm's market orientation. For example, the ability to adapt and respond to the evolving needs of customers is critical to business success in volatile business environment. In such conditions, company executives may develop externally focused activities to identify and fulfil customer needs in addition to monitoring the competition. Moreover, it has been found that perceived environmental turbulence is positively related to a firms' level of customer orientation. This is due to a firm's desire to minimize uncertainty (Davis et al., 1991). On the contrary, in a stable environment where customer types and preferences do not change frequently over time, little adaptation to the marketing mix is required to satisfy customer needs, resulting in a lower degree of customer orientation.

Because market turbulence implies changing strategies in the face of changing customer needs, I expect a negative relationship with performance. Hence, the hypothesis to be tested is:

> *H2a:* The lesser the extent of Market Turbulence, the greater the positive impact of Market Orientation on Business Performance.

Technological Turbulence as a Moderator

Technological turbulence has been defined as the magnitude of change in technology associated with products, services and R&D (Bennet and Cooper, 1981; Miller, 1987). Kohli and Jaworski (1990) define technology as the entire process of transforming input to output and the delivery of the output to the customers. It is argued that the market orientation is more important in industry which is technologically stable than in an industry which is changing rapidly.

In the environment, which is characterized by rapidly changing technologies, technological innovation may be one of the ways of achieving competitive advantage (Covin and Slevin, 1989). Jaworski and Kohli (1993) highlight that for businesses, which work with nascent technologies and experience rapid transformation, the ability to gain competitive advantage is likely to be reduced. On the other hand, the businesses, which operate under stable conditions may rely more on their ability to be market-oriented. Stable conditions allow a long-term allocation of resources to areas such as R&D, which may lead to product improvement that might have a positive impact on business performance. Research and development projects can also increase the efficiency of companies and lower their operating costs. Hayes and Wheelright (1984) report that in a technologically turbulent environment, the majority of innovations and R&D efforts are undertaken outside the company. This phenomenon greatly reduces the amount of research work undertaken, thus reducing the likelihood of becoming market oriented. Further, high rate of technological change may require increased investment in R&D and production capacity (Hayes and Abernathy, 1980; Hayes et al., 1984). Certainly, under such conditions, companies may be reluctant to embrace the market orientation philosophy because the cost and time involved to become market-oriented does not pay enough benefits for a business to be commercially viable in the long-term.

Therefore, it is proposed that there is a negative relationship between technological turbulence and business performance. Hence, the hypothesis to be tested is:

> *H2b:* The lesser the extent of Technological Turbulence, the greater the positive impact of Market Orientation on Business performance.

Competitive Intensity as a Moderator

Competitive intensity has been defined as the changes in competitors' hostility and the extent to which competitors' operations have changed (Narver and Slater, 1990). Competitive intensity involves the aggressiveness of competitors' actions. A hostile environment is characterized by competitors who attack each other aggressively on numerous dimensions, e.g. pricing, promotions, product development, distribution, etc. In a market where competition is stable, a business pays close attention to competitors' costs and strategies, as it is easy to uncover competitors' weaknesses. In turn, it helps gain competitive advantage by developing an appropriate strategy. On the contrary, it is difficult to monitor competitors' actions when competitive intensity is not stable and competitors move back and forth between strategic groups (Porter, 1980).

In highly competitive business environments, customers tend to be faced with several different options to satisfy their needs and wants. In such conditions, there is a tendency for firms to become more sensitive and responsive to the needs of customers (Lusch and Laczniak, 1987). Thus, a business that is not market-oriented is in peril of losing customers to competitors, causing market orientation to become an important determinant of business performance. Covin and Slevin (1989) argue that a strong competition drives companies to seek new products, services and markets to enable them to survive. Therefore, it is imperative that companies subscribe to the market orientation focus. Conversely, in environments where demand exceeds supply, there is a likelihood that firms can get away with low levels of market orientation (Kohli and Jaworski, 1990).

Hence, a positive relationship between competitive intensity and performance is expected. Therefore, the proposed hypothesis to be tested is:

> *H2c:* The greater the degree of Competitive Intensity, the greater the positive impact of Market Orientation on Business Performance.

Recent Developments in the Market Orientation Theory

A review of the literature relating to market orientation and business performance is presented in the following section in chronological order.

Kohli and Jaworski (1990) developed the theory of market orientation after conducting extensive field interviews with 47 US business managers. The conclusion was that the greater the level of market orientation of an organization, the better its business performance. This conclusion is based on the premise that a market orientation appears to provide a unifying focus for the efforts of individuals and departments within an organization, thereby leading to superior performance. Market orientation is about generation and dissemination of market intelligence which is related to customers' current and future needs. Another conclusion was that the market orientation and business performance relationship might be moderated by exogenous factors, such as competition, markets, governments, etc. The degree of variation in the relationship will depend upon the extent to which there are changes in the exogenous factors.

Narver and Slater (1990) conducted a survey in the 140 SBUs of a major western corporation in the US and identified a positive association between market orientation and profitability. *Return on Assets* was used as a business performance indicator. A SBU was defined as an organization with a defined business strategy and a manager with sales and profit responsibility (Aaker, 1980). Each member of the top management team received a questionnaire titled Business Practice Survey, which contained questions relating to companies practices, strategies, competitive environment and performance of the SBU in its principal served market. The response rate was 84 per cent. The sample consisted of two types of businesses: commodity business, such as selling of lumber plywood, wood chips and logs; and non-commodity business, such as hardwood cabinets, laminated looms, oriented strand board and particle board. The study was industry-specific as it relied only on one corporation.

Ruekert (1992) found that higher performing SBUs were found to have a higher level of market orientation than lower performing SBUs. The relationship was tested within a large Fortune 500 high technology company based in the USA. At the time of the study, the company had recently changed its top management team and the chief executive officer had embarked on a programme to assess and improve the company's relationship with its market. The company consisted of five SBUs that had a sales volume of about $3 billion with about 34,000 employees worldwide. These SBUs provided their customers with a wide variety of products and services in the area of computer and information management. The study used multiple respondent survey technique from each SBU to obtain information. The purpose of using this methodology was to obtain the most comprehensive view of the activities of the business while diminishing the impact of individual bias.

Diamantopoulos and Hart (1993) found weak evidence of association between market orientation and *business performance*. The data used in the study were from an earlier survey of companies that were above and below average performers in both high and low growth industries. The performance indicators were profitability and sales growth of companies. Data were collected by means of personal interviews. A response rate of 46 per cent was achieved. The sectors which were above average on *sales growth*, *profit margin*, *return on investment* during the period 1981-1984 in the UK were electronics, medical equipment, food processing and pharmaceutical, whereas sectors which scored below average on the above mentioned indicators were agriculture equipment, paper board, sports and toys.

Jaworski and Kohli (1993) used five indicators of performance: market share, *return on equity*, *organizational commitment*, *esprit de corps* and overall performance in the two-sample study. In both samples, market orientation was not found to be associated with either market share or return on equity; however, it was significantly and positively related to organizational commitment, esprit de corps and overall performance. The first sample was drawn from the member companies of the Marketing Science Institute (MSI) and the second sample was drawn from the top 1000 companies (in terms of sales) listed in the Dun and Bradstreet Million Dollar Directory. A multiple informant design was employed in the sample. A copy of the questionnaire together with a personalised letter and a return envelope was mailed to the marketing executives and non-marketing executives in each SBUs listed in the MSI. The response rate was 89 per cent for the marketing executives and 78 per cent for the non-marketing executives. From the Dun and Bradstreet sampling frame, researchers requested informants to complete and return the questionnaire according to the procedure described for the MSI companies. The response rate was 80 per cent from the marketing executives and 70 per cent from the non-marketing executives. The study was not industry specific as the 500 companies were randomly selected from the top 1000 companies.

Orelowitz (1993) investigated whether the Narver and Slater's (1990) market orientation scale was reliable in the South African business environment. It appeared that the scale was unreliable and that the factors, which emerged from this analysis, differed from Narver and Slater's (1990) study. Further, the instrument failed to yield a positive relationship between market orientation and profitability. The findings of this study raised questions as to the validity of the scale, which was being perceived as an important scale for market orientation.

Chang and Chen (1994) examined the impact of market orientation on the *total offering quality* and business *profitability* by using a sample of 116 security brokerage firms in Taiwan. The extent to which a company offered total offering quality to its customer was measured as the degree of perceived gap between the firms' management capability to serve customers better and the customers expectation to be served better. The narrower the gap, the better customers were served, thus better total offering quality to the customer by the company. The study suggested that a market-oriented business improved the total offering quality which in turn led to superior business results. The market orientation was found to be positively associated with business profitability.

Slater and Narver (1994) used three performance indicators: *return on assets*, *sales growth* and *new product success* in the study. A positive association was found between market orientation and each of the performance indicators. The sample consisted of 81 SBUs in a forest product company and 36 SBUs in a diversified manufacturing corporation, both of which are listed among the Fortune 500 firms. The 81 SBUs of forest-related products are a subset of the sample from the study described in Narver and Slater (1990). The criterion used to select the subset was to exclude those SBUs which primarily utilized telemarketing as their selling approach. The telemarketing selling technique was not considered suitable for the measurement of market orientation as it eliminated the need for contacting the customer in person. Within each SBU, the Top Management Team was identified and each member was sent a questionnaire titled Business Practices Survey containing questions regarding the SBUs' competitive practices, strategies, competitive environment and performance in its principal market. The response rate was 84 per cent in the forest products corporation and 74 per cent in the diversified manufacturing corporation.

Atuahene-Gima (1995) investigated whether the relationship between market orientation and *new product performance* depended upon the degree of product newness to customersand to firms. The study was a cross-sectional survey of a sample of 600 firms comprising services and manufacturing firms, drawn from the Dun and Bradstreet directory of Australia. A response rate of 48 per cent was achieved. The results showed a positive relationship between market orientation and a new product's market performance, proficiency of pre-development activity, proficiency of launch activity, service quality, product advantage, marketing synergy and teamwork. It was also found that market orientation had a greater effect on performance during the early stage of the product life cycle and when the perceived intensity of market competition and industry hostility was high.

Caruana (1995) used return on capital employed, sales growth and overall performance as performance indicators to test the impact of market orientation on performance. A significant positive link between market orientation and business performance was detected. The study utilized Kohli and Jaworski's (1993) market orientation scale and a Likert scale consisting of three items measuring business performance. Two of these items sought to measure *return on capital employed* and *sales growth* of the respondents' firm in the last five years relative to other companies in the industry while the third item asked respondents for their impression of their firms' overall performance in the last five years relative to others in the industry. Personal interviews were conducted with 200 marketing managers and officers responsible for marketing across different types and sizes of firms in Malta. A sample was drawn from a government register. A response rate of 97 per cent was obtained.

Greenley (1995) used three performance indicators: *return on investment*, *new product success* and *sales growth* to test the effect of market orientation on business performance. No significant positive association was found between market orientation and any of the above mentioned performance indicators; however, a moderating effect from external environmental variables on the market orientation - business performance relationship was detected. Narver and Slater's (1990) market orientation scale, after making some minor changes to make it compatible with the UK business culture, a questionnaire and a personal letter were mailed to the managing directors or chief executive officers of 1,000 companies. The sample was randomly drawn from the Dun and Bradstreet database of UK companies with more than 5,000 employees. A total response of 280 was obtained. The sample had a balanced proportion of consumer, industrial, service and product companies.

Atuahene-Gima (1996) examined the influence of market orientation on innovation characteristics in sample of 158 manufacturing and 147 service firms in Australia. The results indicated that market orientation had a significant relationship with innovation characteristics, such as *innovation-marketing fit*, *product advantage* and *interfunctional teamwork* but not with the *product newness* and *innovation-technology fit*. Further, after controlling the effect of these innovation characteristics, it was found that in both the product and service innovation samples, market orientation significantly contributed to the innovative project's performance. These results did not confirm the hypothesis that market orientation had stronger impact on service innovation than on product innovation.

Bhuian (1996) found Organizational Performance of Saudi manufacturing companies to be a positive function of market orientation. Data were collected through postal surveys in two industrial cities in Saudi Arabia. Market orientation was measured on the scale adapted from Kohli and Jaworski's (1993) study. A response rate of 77 per cent was achieved.

Langerak (1996) studied the relationship between market orientation and *competitive environment*, *strategy* and *competitive advantage*. The central proposition was that the competitive environment and the attitude of top management determined a firm's strategy. The latter affected a firm's market orientation, thus a competitive position in the marketplace. The sample consisted of 550 small- and medium-sized industrial firms in the Netherlands. A postal survey was used to collect the data. The response rate was 23 per cent. The study used the multi-item scales for the measurement of market turbulence (Moriarty and Thomas, 1989), strategic orientation (Venkatraman, 1989), management attitude (Kohli and Jaworski, 1990), market orientation (Narver and Slater, 1990) and competitive advantage (Dess and Davis 1984; Day and Wensley, 1988). Results indicated that competitor orientation was the primary intermediating component of market orientation explaining the competitive advantage of small- and medium-sized firms. Customer orientation and interfunctional co-ordination were found to be of secondary importance. A direct effect of the entrepreneurship on market orientation and competitive advantage was also found.

Pitt et al. (1996) used *return on capital employed*, *sales growth* and *overall performance* to test the effect of market orientation on performance. A positive association between market orientation and business performance was found. Questionnaires were mailed to the top 1000 companies (in terms of numbers of employees in the last three years), drawn from the FAME CD-Rom database. The response rate was 17 per cent.

Pelham and Wilson (1996) found a significant positive correlation between market orientation and *relative product quality*, *new product success* and *profitability*; however, no significant correlation between market orientation and sales growth and market share were detected. The study differed from the previous studies in two ways: First, this study was conducted in small firms in the USA. The range of firm sizes was 15 to 65, reducing the likelihood of the impact of firm size on performance. Second, this was a two-year longitudinal study. The sample was drawn from the Centre for Entrepreneurship, Eastern Michigan University, USA, which consisted of 370 Michigan firms with average sales about $3 million. A respondent base of 68 firms was used.

Appiah-Adu (1997) examined whether the market orientation-business performance link established in large firms were applicable in SMEs. Further, the study tested the possible effects of *market growth*, *competitive intensity*, *market turbulence* and *technological turbulence* on the market orientation-business performance relationship. Performance measurement indicators were: *new product success*, *sales growth* and *return on investment*. Pelham and Wilson's (1996) market orientation scale was adapted for the purpose of this study because of its applicability to small business. Questionnaires were mailed to the managing directors of 500 small firms, which were randomly drawn from the Dun and Bradstreet database of UK firms with employees between 10 and 50. A response rate of 20 per cent was achieved. The finding indicated that there was a positive impact of market orientation on business performance; however, mixed effects of the competitive environment on the market orientation - performance relationship were observed.

Appiah-Adu (1998) investigated the link between market orientation and business performance in the transition economy of Ghana. Potential influences of market dynamism, competitive intensity and market growth on the relationship were examined. The market orientation scale was adapted from Golden et al.'s (1995) and Narver and Slater's (1990) studies. The sample comprised 200 large organizations, which were randomly selected from the Ghanian Business Directory (1996). A response rate of 37 per cent was obtained. The results indicated that market orientation did not appear to have a direct impact on sales growth and return on investment, but the competitive environment influenced the market orientation-business performance link.

Appiah-Adu and Singh (1998) investigated into two key areas in the SMEs: first, the effect of market orientation on business performance; and second, the moderating influences of *market dynamism*, *competitive hostility* and *market growth* on the market orientation-business performance link. Selected firms consisted of 500 SMEs, drawn randomly from the FAME CD-ROM Database. A response rate of 26 per cent was achieved. Findings suggested that market orientation influenced certain dimensions of SMEs performance and that competitive environment moderated the relationship between market orientation and business performance.

Loubser (2000) defined market orientation as business culture that focused on creating mutually rewarding relationships between customers and the organization based on the interest of *stakeholders*, *organizational competitive advantage* and *core competencies*. The study took a systematic view of market orientation rather than cause and effect view. Data were collected from 449 unlisted and 51 listed organizations in South Africa. Seven performance indicators were used in the study: growth in *market capitalization*, *growth in total assets*, *return on equity*, *return on assets*, *growth in sales* and *price earnings* – organization versus sector. The study concluded that only one performance measurement – return on equity – was significantly correlated with market orientation in its consolidated form; however, business behavior was significantly correlated with two performance measurements – growth in total assets and return on equity. Further, external variables, such as technology change, price sensitivity, market growth, competitive intensity and others were not significantly correlated with the market orientation.

Sin et al. (2000) investigated into the link between market orientation and business performance in the transition economy of China which was transforming itself from a planned economy to a market economy. A total of 1200 companies based in Beijing were randomly selected from the Beijing Yellow pages Commercial/Industrial Telephone Directory. Telephone calls were made to the top administrator of these companies to solicit their participation in the survey. Administrators of the 300 companies, which agreed to take part in the study, were contacted in person for the purpose of data collection. The questionnaire was administered in Chinese. Altogether, 210 completed questionnaires were returned, representing a response rate of about 18 per cent. The study found that market orientation was positively and significantly associated with *sales growth*, *customer retention* and *overall performance*.

Grewal and Tansuhaj (2001) investigated the contingent nature of the influence of market orientation and strategic flexibility on firm performance after a crisis has occurred. Data were collected from Thai SMEs in three waves. First wave consisted of 49 middle managers and owners participating in an executive MBA program at a large university in northeastern Thailand. A subsequent group of respondents participating in the same program was a set of 61 respondents. Third wave comprised of data collection from 22 firms affiliated with the conglomerate. Of 132 responses, 120 were complete and usable questionnaires. The questionnaires were translated from original English version to Thai, and then used the back-translation technique to ensure that the original meaning was maintained. The results of the study indicated that market orientation had an adverse effect on firm performance after a crisis and that this effect was moderated by *demand* and *technological uncertainty*, and was enhanced by *competitive intensity*. In contrast, strategic flexibility had a positive influence on firm performance after a crisis, which was enhanced by *competitive intensity* and moderated by *demand* and *technological uncertainty*. In conclusion, it appeared that market orientation and strategic flexibility complemented each other in their efficacy to help firms manage varying environmental conditions.

Soehadi et al. (2001) developed a market orientation scale in the context of retail industry and tested its impact on *supplier partnership* and *retail performance* of firms. Data were collected from 172 Indonesian retail firms, of which 159 questionnaires were complete and usable. A response rate of 36.5 per cent was achieved. The questionnaire was translated into Indonesian and the quality of the translation was subsequently verified using back translation by independent judges. The findings of the study showed that market orientation had positive effects on both supplier partnership and retail performance.

Vazquez et al. (2001) examined the psychometric properties of the newly developed market orientation scale, and then the relationships between market orientation and the following variables were tested: *firms' commitment to the innovation activities*, *effective innovation rates*, *degree of innovativeness of the new products developed*, *firms' competitive strategy* and *companies' performance*. A total of 264 companies were drawn as a sample from the Directory of Industrial Companies and Industrial support Services in the Principality of Asturias, with the aim to include a broad range of Spanish companies which varied by sizes and industry types. The data were collected by means of personal interviews with the directors of these companies. Altogether, 174 companies agreed to participate in the study. The results indicated that market orientation had a direct and positive effect on a firm's innovativeness, but the direct significant effect of market orientation on the company's innovation rate and performance was not detected. Further, market orientation was found to have an indirect effect on companies' innovation rate and performance, which were mediated by firm's innovativeness, innovation rates and new product innovativeness.

Gainer and Pandanyi (2002) examined the relationship between market orientation and organizational performance in the context of nonprofit art organizations based in Canada. The three indicators were used to measure organizational performance: *customer satisfaction*, *peer reputation* and *resource attraction*. Data were collected by sending a ten-page, 146-question questionnaire to the General Managers or CEOs of 705 nonprofit arts and cultural organizations in the Greater Toronto and Greater Montreal areas in Canada. The sample list was drawn from: lists of 1998 and 1999 grant recipients provided by arts councils at the local, provincial and federal levels; and the Directory of Canadian Cultural World Wide Web sites available at the Cultural website. A total of 650 qualified organizations were contacted, of which 138 organizations returned complete questionnaires that were used in the analysis. A final response rate of 21.2 per cent was achieved. The questionnaires were in French and English. The study concluded that market-oriented culture was significantly and positively related to growth in resources, customer satisfaction and growth in reputation among peers.

Matsuno et al. (2002) investigated both direct and indirect influences of *entrepreneurial proclivity* and *market orientation* on *business performance*. For data collection, a master list of 1300 US manufacturing companies that identified one marketing executive (vice president or director level) per company from a well-known, Midwest-based commercial vendor was obtained. A questionnaire along with cover letter and stamped return envelope were sent to a random sample of 1000 executives of the 1300 companies in the master list. Three-wave mailings produced an effective response rate of 38.7 per cent (or 364 usable responses). The results indicated that entrepreneurial proclivity had a positive and direct relationship with market orientation, and that it had an indirect and positive effect on market orientation through the reduction of departmentalization. Further, the results suggested that entrepreneurial proclivity's performance influence was positive when mediated by market orientation but negative or nonsignificant when not mediated by market orientation.

Tay and Morgan (2002) tested the theory of market orientation in the context of chartered surveying industry in the UK. Following tailored design methodology, data were collected in a mail survey from 179 general practice chartered surveying firms, the largest of the seven divisions within the Royal Institution of Chartered Surveyors (RICS). Respondent firms seemed to be representative of the population of UK general practice surveying firms. A response rate of 24 per cent was achieved. The study indicated that market orientation had a positive impact on a firm's business and marketing performance. Further, it was found that firms with more *risk tolerant* senior managers and more *formalized* and *specialized* marketing organization structures had higher levels of market orientation. The study confirms that the relationship between market orientation and business performance is robust across different environments.

Noble and Sinha (2002) explored: the relative performance effects of various dimensions of market orientation using a longitudinal approach based on letters to shareholders in corporate annual reports; the relative effects of alternative strategic orientations that reflected different managerial priorities for the firm; and the mediating effects of organizational learning and innovativeness on the orientation-performance relationship. Annual reports of companies were analyzed by using cognitive mapping technique, which converted texts of the reports into quantitative data by coding each sentence of the letters for the firms and years studied. The results showed that firms possessing higher levels of competitor orientation, national brand focus, and selling orientation exhibited superior performance.

Delbaere et al. (2003) tested the moderating and mediating effects of knowledge management on market orientation-business performance link. Performance indicators used were *customer loyalty* and *financial performances*. Data were collected from 165 managers across Canada using a web survey. Respondents were top- or middle-level managers from different functional areas of at least 28 companies in several industries. The study utilized multi-informant methodology for the collection of data. The findings of the study suggested that market orientation did not have a direct impact on financial performance of companies, that having a market orientation generally resulted in greater inclination to engage in knowledge management, which in turn led to higher customer loyalty for the company, and that greater knowledge led to higher customer loyalty, which in turn led to higher financial returns.

Singh (2003) ascertained the extent to which Indian industrial firms had adopted the concept of market orientation, as a result of economic reform, by examining the link between the market orientation and *business performance*. The potential effects of *competitive intensity* and *market dynamism* were also examined. Data for this study were collected by administering questionnaires in person, following a pre-test with a group of Indian executives in three major cities (New Delhi, Bombay and Calcutta) in India. For the selection of the respondents, only indigenous firms with a formal marketing/sales department were included in the survey. The initial list comprised of firms in the three major cities that were listed in the Kompass Indian Business Directory (2000). From the initial list of firms, which included the name of CEOs and senior marketing executives, sixty executives in each of the cities were randomly selected for the survey. A total response of 138 was received. Findings of the study suggested that there was a positive change in the market orientation and that the market orientation was significantly and positively related to return on investment and customer retention. Further analysis indicated that for a high level of competitive intensity and a low level of market dynamism, the relationship between the market orientation and performance was stronger.

Conclusions

This chapter built a foundation for systematic development of the theory of market orientation by discussing the three conceptualizations of Kohli and Jaworski (1990) and Narver and Slater (1990) and Deng and Dart (1994). For purpose of the study, market orientation was defined as the 'set of activities co-ordinated in such a way that derives customer satisfaction through superior performance of products (machines) and related services (training, maintenance, etc.) while still being competitive (price, responsiveness, delivery, etc.) in the marketplace'.

As the objective of the chapter was to construct the theory, theoretical arguments were made that the degree of market orientation was positively related to business performance as measured by an efficiency-based indicator, *return on investment*, effectiveness-based indicators, *sales growth* and *market share*, adaptability-based indicators, *new product success rate*. The study utilized two additional performance indicators: customer-satisfaction based indicator, *customer retention* and export-based indicator, *global presence*. Further, it was postulated that market orientation was significantly related to *return on investment* when *market turbulence* is low, when *competitive intensity* was high and when *technological turbulence* was low.

A parsimonious review of selected studies pertaining to market orientation is presented at the end of the chapter. Although there was little agreement on the generalizability of the findings of these studies, it was interesting to note that sector-specific studies seemed to yield a positive relationship between market orientation and business performance whereas cross-sectional studies appeared to offer a mixed support for the relationship.

In the next chapter, an attempt is made to clarify the domain of the market orientation and provide a foundation for developing the market orientation scale. Reliability and validity of the scale are also tested.

Chapter 3

Preliminary Empirical Analysis

Introduction

In the last few decades, there has been much development in the art and science of marketing research. This development has been primarily in terms of research methodologies from other disciplines and an increased sophistication in the use of both hardware and software for data collection, analysis and presentation. However, industrial marketing research is an area where little work has been done to develop and test sector-specific scales.

The chapter is concerned with the development of the market orientation scale. The paradigm suggested by Churchill (1979) is followed and described in detail. The comparative analysis among these existing market orientation scales (Narver and Slater, 1990; Jaworski and Kohli, 1993; Deng and Dart, 1994) has been performed for the generation of the scale items. Data were collected from the machine tool industry in the category of SIC 3541 and SIC 3542 by using single-informant postal survey methodology. Two-wave mailing yielded an effective response rate of 24 per cent. Preliminary results indicated that the 35-item market orientation scale is represented by a seven-factor solution, explaining 67 per cent of the variance in the scale. These factors were labelled as customer focus, competitor focus, responsiveness, customer satisfaction focus, market information, marketing focus and negligence focus. The reliability of each factor was measured by Cronbach Alpha and split-half Alpha methodology. Finally, some aspects of validity of the scale were examined.

The Paradigm and the Methodology

To strengthen the design of the study, the triangulation methodology was employed. This meant the utilisation of several data and the use of both quantitative (positivism) and qualitative (phenomenological) approaches (paradigms). Filstead (1979) defines a paradigm as a positional approach which:

* Serves as a guide to the professionals in a discipline, which indicates what are the important problems and issues confronting the discipline.
* Develops explanatory models and theories, which can place problems in a framework, which will allow practitioners to solve them.
* Establishes the criteria for the appropriate tools, such as methodologies, instruments and types and forms of data collection, to use in solving disciplinary puzzles.
* Provides an epistemology in which the preceding tasks can be viewed as

organising principles for carrying out the normal work of the discipline. Paradigms not only allow a discipline to make sense of different kinds of phenomena but also provide a framework in which these phenomena can be identified as existing in the first place.

Although there is a growing acceptance of qualitative methods in social science, there has been a debate as to what relationship should be between qualitative and quantitative methods. There are two groups: the purists and the pragmatists (Rossman and Wilson, 1985). The purists believe that these two methods are not compatible because they are based on paradigms that make different assumptions about the world and about what constitutes a valid research (Guba, 1978). They seek logic of justification for the method used that begins with the basic principles about truth, reality and the relationship between the investigator and investigated research objectives (Smith and Heshusius, 1986). So, there is a relationship between the paradigm and the select method. On the other hand, pragmatists see a more instrumental relationship between paradigms and the selected method. To them, methods are collections of techniques. Hence, the attributes of a paradigm are not inherently linked to either qualitative or quantitative methods. Both methods types can be associated with either the attributes of qualitative or quantitative paradigms (Reichard and Cook, 1979). Indeed, pragmatists have gone on to combine the methods in practice (Smith and Louis, 1982).

A survey of the literature indicates that different methodologies can be combined fruitfully. There is an association between the method type and paradigm (Reichard and Cook, 1979). The association has to do with the means of expression rather than logic (Eisner, 1981). Quantitative studies are usually based on a positivist paradigm which assumes that there are social facts with an objective reality, apart from the beliefs of individuals, while qualitative research is based on phenomenological beliefs which holds the fact that reality is socially constructed through individual or collective definitions of the situations (Taylor and Bogdan, 1984). Quantitative research seeks to explain the causes of changes of social *facts* primarily through objective measurement and quantitative analysis, whereas qualitative research is more concerned with the understanding of the social phenomenon from the actors' perspectives through participation in the life of those actors (Taylors and Bogdan, 1984).

Denzin (1978) identified four basic types of triangulation methodology:

* Data triangulation is about the use of a variety of data sources in a study and is based on the assumption that understanding a social phenomenon requires its examination under a variety of conditions.
* Investigator triangulation involves the use of several different researchers or evaluators in the research process; however, the decision about who these multiple researchers should be and what their roles should be in the research process is problematic.
* Theory triangulation is about the use of multiple perspectives to interpret a single set of data.
* Methodological triangulation is the most commonly used triangulation and it refers to the use of multiple methods in the examination of a single social phenomenon.

The research design of this study is based on the methodological triangulation methodology which uses multiple methods and sources of data. Triangulation is perceived to be a strategy for improving the validity of the research and is supposed to support a finding by showing that independent measures of it agree with it or at least do not contradict it (Miles and Huberman, 1984). It is essentially a strategy that aids the elimination of bias and allows the dismissal of plausible rival explanations (Denzin, 1978).

This study starts with the positivist paradigm, which holds that behaviour can be explained through objective facts. It also assumes that these objective facts can be measured and that bias and error can be reduced in the measurement instrument. In the following section there is a discussion on item generation, purification and redevelopment of the market orientation scale by using a factor, reliability and validity analyses.

In chapter six, qualitative analysis of four mini case studies is presented. This approach was followed by in-depth interviews with the top managers in select four companies with the view to triangulate qualitative and quantitative methodologies. It was important to triangulate because different methods could produce a different understanding of the market orientation and its effect on performance. It was assumed that by using qualitative methods, the bias inherent in any data source, investigator and method would be reduced when used in conjunction with other investigative methods. It is believed that when triangulation is used as a research strategy, the result will converge upon the truth about some social phenomenon.

The purpose of the study is to redevelop a market orientation scale by conducting a comparative analysis of three market orientation scales developed by Narver and Slater (1990), Jaworski and Kohli (1993) and Deng and Dart (1994). These authors generated a pool of items following an extensive literature review and field research. Following the comparative analysis of these scales, a new redeveloped scale was presented which was more comprehensive in nature and appeared to tap overall activities relating to market orientation in the manufacturing industry. The need to redevelop the scale was important because these scales had an overlap on a conceptual and operational basis, so these items needed filtering while keeping in mind the new definition of market orientation (p. 19) as the focus of the scale. If a focal construct is not adequately defined, it is difficult to develop a scale that faithfully represents its domain. Further, failure to clearly define the focus of the scale makes it difficult to correctly specify how the scale should be related to its measures. Bagozzi and Fornell (1982) and Bollen and Lennox (1991) have presented two different types of measurement indicators: *reflective indicator* measurement models posit that the direction of causality flows from the scale to the measures, whereas *formative indicator* measurement models posit that causality flows from the scale to the scale. By having a clear, concise conceptual definition of the market orientation, it was determined that reflective indicator measurement was the most appropriate for the development of the scale. Clearly, this was important because measurement model misspecification could undermine both construct validity and statistical conclusions.

In the following sections, attempts were made to employ qualitative and quantitative methodologies to have greater confidence in the development of the scale and research findings.

The Procedure

Domain Specification and Definition

Psychologists were one of the first social scientists to develop methods for constructing scales to measure behavioural variables (Ghiselli, 1964; Likert, 1967; Nunnally, 1978). The procedure to develop the market orientation scale largely followed the guideline recommended by Churchill (1979), which is illustrated in Figure 3.1.

The first step entailed the specification of domain of the market orientation, which appeared to be complex as there was no single definition of the philosophy of market orientation. The review of the literature revealed that there were a number of meanings ascribed to market orientation. For example, Konopa and Calabro (1971) placed a greater emphasis on customer than production and cost-related activities, whereas Felton (1959) and McNamara (1972) suggested that the involvement of marketing executives in strategic decision making process and the integration of activities within the marketing function was the central issue for a company to be market oriented.

Narver and Slater (1990) conceptualised market orientation as a three-component scale: customer orientation, competitor orientation and interfunctional co-ordination. On the other hand, Kohli and et al., (1993) have identified intelligence generation, intelligence dissemination and responsiveness as the basic components of market orientation. Although these authors differed in their preferred conceptualisations, it is evident that there are two main themes underlying these definitions: customer and competitor focus, i.e. generation of information pertaining to customers and competitors; and responsiveness, i.e. dissemination of the information across functional departments. Deng and Dart (1994) have the similar concept of market orientation to Narver and Slater (1990) but assume profit emphasis as one of the components of market orientation. This study takes the view of the combined components of market orientation definitions given by these authors to specify the domain of the market orientation scale except profit emphasis as I believe that profit is the outcome of the adoption of the marketing concept and not the goal of the concept. (Levitt, 1969).

Item Generation

Cadogan and Diamantopoulos (1995) conducted a comparative analysis between the components of Kohli and Jaworski's (1993) and Narver and Slater's (1990) market orientation scales and have shown the conceptual and operational overlap between these scales.

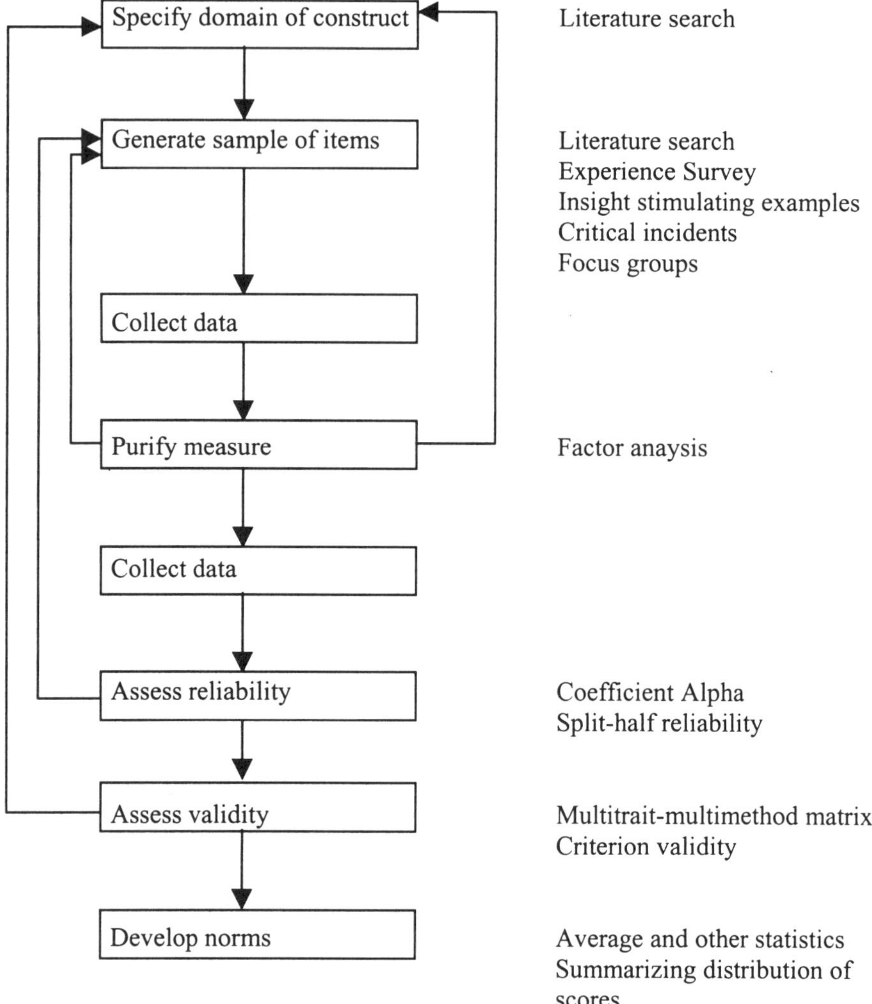

Figure 3.1 Churchill's guideline for scale development

Source: Churchill (1979)

Figure 3.2 presents the framework for the comparative analysis which was performed among these market orientation scales (Narver and Slater, 1990; Jaworski and Kohli, 1993; Deng and Dart, 1994) with the following aims:

* Detect the overlap among these scales so the duplication of items could be deleted and new items tapping the market orientation domain could be added.

* Generate a comprehensive list of items representing the market orientation,
* Purify the market orientation scale by employing factor and reliability analyses.

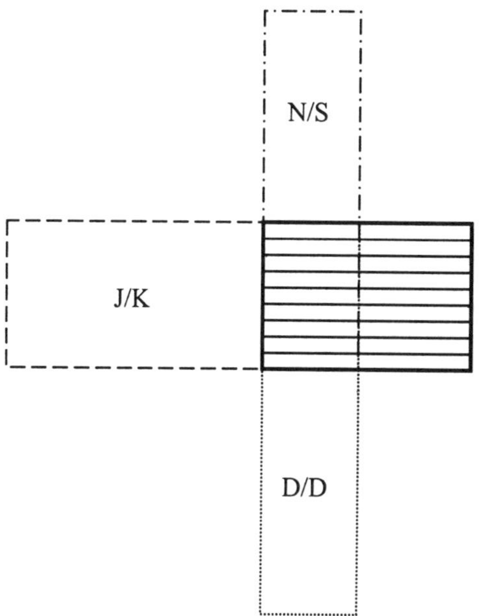

→ The author's market orientation scale

J/K=Jaworski and Kohli (1993)
N/S=Narver and Slater (1990)
D/D (1994)= Deng and Dart

Figure 3.2 The conceptual framework for the comparative analysis

Source: The author

After comparing the customer orientation dimension of Narver and Slater's (1990) scale with the equivalent dimension of other market orientation scales, it was revealed that there were items that were not only associated with customer orientation but also associated with customer satisfaction. As the objective of the study was to develop a comprehensive scale that would tap the domain of market orientation in the context of machine tool industry, an additional conceptual dimension of market orientation – customer satisfaction – was introduced. Tables 3.1a through 3.1d identified items having the same shade of meaning from each of the market orientation scales.

Customer orientation: Narver and Slater's (1990) customer orientation refers to sufficient understanding of a firm's target buyers to be able to create superior value. This statement has a conceptual overlap with Kohli and Jaworski's (1990) intelligence generation dimension of the market orientation scale. Deng and Dart (1994) have generated items, such as 'we measure customer orientation on a regular basis' (Table 3.1a), 'we encourage customer comments - even complaints - because they help us to do better' (Table 3.1b), 'after sales service is an important part of our business strategy' (Table 3.1c), 'we have strong commitments to our customers' (Table 3.1d) to tap the concept of market orientation. Remaining items from Deng and Dart's (1994) market orientation scale, such as 'we concentrate on production and let our distributors worry about sales,' 'our company would be much better off if our sales force just worked harder,' 'for our customers, price is the most important selling feature,' were not selected for the redevelopment of the scale, as they perceived to be related to production and sales orientations and were found to be least representative of the machine tool industry. This was confirmed through face-to-face interview with managers. On the other hand, items, such as 'in our company, marketing department's most important job is to promote our product and services to our customers,' and 'in our company, marketing's most important job is to identify and help meet the needs of our customers' were selected with the aim to check the extent to which the role of the marketing function contributed to the success of a company undergoing through a process of market orientation. Kohli and Jaworski (1993) taped the role of marketing by asking managers the extent to which there was a degree of communications between marketing and manufacturing departments. This item was retained in the scale. Finally, none of the six items was selected from Narver and Slater's customer orientation dimension, as the main theme of these items was captured either by Kohli and Jaworski's or Deng and Dart items.

Table 3.1a Customer orientation vs. intelligence generation

Conceptual overlap: Yes. Operational overlap: Yes. Operational examples:
KJ: We poll end-users at least once a year to assess the quality of our products and services.
NS: Measure customer satisfaction.
DD: We measure customer satisfaction on a regular basis.

Table 3.1b Customer orientation vs. intelligence dissemination

Conceptual overlap: Yes. Operational overlap: Ambiguous. Operational examples:
KJ: Data on customer satisfaction are disseminated at all levels in this business unit on a regular basis.
NS: Understand customer needs.
DD: We encourage customer comments - even complaints - because they help us to do a better job.

Table 3.1c Customer orientation vs. responsiveness

Conceptual overlap: Yes. Operational overlap: Ambiguous. Operational examples:
KJ: For one reason or another, we tend to ignore changes in our customer's product or service needs.
NS: Engage in after-sales-service.
DD: After sales service is an important part of our business strategy

Table 3.1d Customer orientation vs. customer satisfaction

Conceptual overlap: Yes. Operational overlap: Yes. Operational examples:
KJ: In this business unit, we meet with customers at least once a year to find out what products or services they will need in the future.
NS: Places emphasis on customer commitment: Customer satisfaction objectives.
DD: We have a strong commitment to our customers.

Competitor orientation: It is the second most important dimension of market orientation. Two items from Narver and Slater's (1990) scale, such as 'top managers discuss competitors' strategies,' and 'companies target opportunities for competitive advantage,' were selected as these items were related to competitive activities. Further, one more item from Kohli and Jaworski's scale, 'several departments get together periodically to plan a response to changes taking place in our business environment' was selected. There was a duplication of an item – we target opportunities based on competitive advantage – between Narver and Slater's (1990) and Deng Dart's (1994) market orientation scales. Detailed analysis of items is presented in Tables 3.2a through 3.4d.

Table 3.2a Competitor orientation vs. intelligence generation

Conceptual overlap: Yes. Operational overlap: Ambiguous. Operational examples:
KJ: In our business unit, intelligence on our competitors is generated independently by several departments.
NS: Salespeople share competitor information.
DD: Our salespeople are instructed to monitor and report on competitive activities.

Table 3.2b Competitor orientation vs. intelligence dissemination

Conceptual overlap: Yes. Operational overlap: Yes. Operational examples:
KJ: When one department finds out something important about competitors, it is slow to alert other departments.
NS: Top managers discuss competitors' strategies.
DD: Our top managers often discuss competitors programs.

Table 3.2c Competitor orientation vs. responsiveness

Conceptual overlap: No. Operational overlap: Yes. Operational examples:
KJ: If a major competitor were to launch an intensive campaign targeted at our customers, we could implement a response immediately.
NS: We respond rapidly to competitors' actions.
DD: We respond rapidly to competitors' actions.

Table 3.2d Competitor orientation vs. customer satisfaction

Conceptual overlap: Yes. Operational overlap: Ambiguous. Operational examples:
KJ: Principles of market segmentation drive new product development efforts in this business unit.
NS: Not applicable.
DD: Not applicable.

Interfunctional co-ordination: Narver and Slater (1990) define interfunctional co-ordination as the coordinated utilization of a company's resources in creating a superior value for customers. Kohli and Jaworski (1993) define responsiveness as the action taken in response to intelligence generation and its dissemination. There is a conceptual overlap; e.g. Kohli and Jaworski's item 'several departments get together to plan a response to changes taking place in our business environment' and Narver and Slater's item 'engage in interfunctional integration of strategy' are conceptually similar. Deng and Dart (1994) used a similar item - we do a good job of integrating the activities of each department - to Narver and Slater's interfunctional co-ordination dimension. Three additional items from Deng and Dart (1994) were selected, as these items represented activities relating to interfunctional co-ordination. They are: 'in our company, the marketing people have a strong input into the development of new products,' 'the marketing people in our organization interact frequently with other departments, such as manufacturing, physical distribution,' and 'in our company marketing is seen as a guiding philosophy for the entire organization'. Further, two items were selected from Narver and Slater's interfunctional co-ordination dimension: 'all functions contribute to customer value,' and 'share resources with other business units'. Detailed analysis of items is presented in Tables 3.3a through 3.3d.

Table 3.3a Interfunctional co-ordination vs. intelligence generation

Conceptual overlap: Yes. Operational overlap: Yes. Operational examples:
JK: Individuals from our manufacturing department interact directly with customers to learn how to serve them better.
NS: Engage in interfunctional customer calls.
DD: People other than our sales representatives (such as top management, research, production) regularly call on customers.

Table 3.3b Interfunctional co-ordination vs. intelligence dissemination

Conceptual overlap: Yes. Operational overlap: Yes. Operational examples:
JK: We have interdepartmental meetings at least once a quarter to discuss market trends and developments.
NS: Share information among functions.
DD: Market information is shared with all departments.

Table 3.3c Interfunctional co-ordination vs. responsiveness

Conceptual overlap: Yes. Operational overlap: Yes. Operational examples:
JK: Several departments get together to plan a response to changes taking place in our business environment.
NS: Engage in interfunctional integration of strategy.
DD: We do a good job of integrating the activities of each department.

Table 3.3d Interfunctional co-ordination vs. customer satisfaction

Conceptual overlap: Yes. Operational overlap: Ambiguous. Operational examples:
JK: When we find out that customers are unhappy with the quality of our service, we take corrective action immediately.
NS: All functions contribute to create customer value.
DD: We look for ways to create customer value in our products.

Purification of Items

While selecting the items for the market orientation scale, care was taken to tap the domain of market orientation as closely as possible. Criteria of uniqueness and the ability to convey different shades of meaning to informants were also used (Churchill, 1979; p. 68). Items, 4, 9, 17, 18, 19, 21, 23, 25 and 29 were reverse-coded to minimize the response set bias.

Since the market orientation scales of Narver and Slater and Jaworski and Kohli are American and Deng and Dart's scale is Canadian, it was important to make these items compatible with the UK business culture. Because of the centrality of the market orientation, each item was critically evaluated for its clarity and appropriateness in personally administered pretests with a panel of five professors. These professors were asked to critique the questionnaire, offer suggestions and indicate the items which were ambiguous in nature or difficult to understand. A 7-point Likert scale was used (1 being strongly disagree and 7 being strongly agree) to enable respondents to indicate the extent to which their companies had adopted the practices described on each of the 45 items. Based on the feedback received from them, it was discovered that items, 1, 3, 5, 12, 18, 19, 26, 33, 36, 37, 38, 39, 40 and 43 needed rephrasing. Two items were eliminated, as they did not seem to relate to the machine tool industry.

Following these pretests, 43 items were retained in the final questionnaire. This pretest was followed by a second phase of pretests by administering postal questionnaire to 30 machine tool manufacturers in the UK. Five completed questionnaires were returned. Very few concerns were raised and only minor refinements were made. For example, it was learnt that some respondents were not familiar with the SIC system of industries, so these codes were expanded in the questionnaire. The inventory of items is presented in Table 3.4.

Table 3.4 Inventory of items relating to market orientation

Information Generation

1 We meet customers at least once a year to find out what product or services they will need in the future. (W/JK)
2 Individuals from our manufacturing department interact directly with customers to learn how to serve them better. (JK)
3 We do a lot of in-house research of our customers' and markets. (W/JK)
4 We are slow to detect changes in our customers' product preferences. (R/JK)
5 We frequently poll end-users to assess the quality of our products and services. (W/JK)
6 We often talk with or survey those who can influence our end-users' purchases, e.g. agents, retailers, distributors. (JK)
7 We collect industry information through informal means, e.g., lunch with industry friends, talks with trade partners, etc. (JK)
8 In our business unit, intelligence on our competitors is generated independently by several departments. (JK)
9 We are slow to detect fundamental shifts in our industry, e.g. competition, technology regulation. (R/JK)
10 We periodically review the likely effect of changes in our business environment (regulation) on customer. (JK)

Information Dissemination

11 A lot of discussion in this business unit concerns our competitors' tactics and strategies. (W/JK)
12 We often have interdepartmental meetings to discuss market trends and developments. (W/JK)
13 Marketing personnel in our business unit spend time discussing customer's future needs with other functional departments. (JK)
14 Our business unit periodically circulates documents (e.g., reports, news letters) that provide information on our customers. (JK)
15 When something important happens to a major customer or market, the whole business unit knows about it in a short period. (JK)
16 Data on customer satisfaction is disseminated at all levels in this business unit on a regular basis. (JK)
17 There is a minimal communication between the marketing and manufacturing departments concerning market development. (R/JK)

18 Departments are slow to disseminate competitor information amongst each other. (W/R/JK)

Response Design

19 It takes us a long time to decide how to respond to our competitor's price change. (W/R/JK)
20 Principles of market segmentation drive new product development efforts in this business unit. (JK)
21 For one reason or another we tend to ignore changes in our customers' product or service needs (R/JK)
22 We periodically review our product development efforts to ensure that they are in line with what customers want. (JK)
23 Our business plans are driven more by technological advances than market research. (R/JK)
24 Several departments get together periodically to plan a response to changes taking place in our business environment. (JK)
25 The product lines we sell depend more on internal politics than real market needs. (R/JK)

Response Implementation

26 We are generally quick to respond to competitor campaigns targeted at our customer base. (W/JK)
27 The activities of different departments in this business unit are well coordinated.
28 Customers complaints fall on deaf ears in this business unit. (R/JK)
29 Even if we came up with a great marketing plan, we probably would not be able to implement it in a timely fashion. (R/JK)
30 We are quick to respond to significant changes in our competitors' pricing structures. (JK)
31 When we find out that customers are unhappy with the quality of services, we take corrective action immediately. (JK)
32 When we find that customers would like us to modify a product or service, the departments involved make concerted efforts to do so. (JK)

Customer Orientation

33 In our company, there is little distinction between sales and marketing. (W/DD)
34 In our company, marketing's most important job is to promote our product and services to our customers. (DD)
35 In our company, marketing's most important job is to identify and help meet the needs of our customers. (DD)

Competitor Orientation

36 In our company, top managers discuss competitors' strategies regularly. (W/NS)
37 The company targets specific opportunities in order to gain competitive advantage. (W/NS)

Interfunctional Co-ordination

38 In our organization, customer calls are disseminated across other departments. (W/NS)

39 In our organization, all departments contribute to create customer value. (W/NS)

40 Within the organization, we share resources with other business units. (W/NS).

41 The marketing people in our organization interact frequently with other departments, such as manufacturing, finance, distribution etc. (DD)

42 In our organization, marketing is seen as a guiding philosophy for the entire organization. (DD)

43 In our organization, the marketing people have a strong input into the development of new products. (W/DD)

W=Reworded, R=Reverse Coded, JK=item from Jaworski and Kohli's market orientation scale (1993), DD=item from Deng and Dart's market orientation scale (1994) and NS=item from Narver and Slater's market orientation scale (1990).

Data Collection

SIC Code: Standard Industrial Classification (SIC) Code is a numbered coding system designed to classify the line of business of a company. For the sake of uniformity across industries, there is a comprehensive US 1972 system being used internationally. There are also British equivalent SIC codes. The SIC code system divides the machine tool industry into two categories: the metal cutting industry (SIC 3541) and the metal forming industry (SIC 3542). Machine tools are power driven machines that cut, form or shape metal. They turn casting, bar-stock, or sheet metal into finished components. The industry operates in numerous markets, identified by the following two-digit SIC categories. They are shown in Table 3.5.

Table 3.5 Two-digit SIC codes

Industry	US SIC 2-digit Code
Agriculture, Forestry and Fishing	01-09
Mining	10-14
Construction	15-17
Manufacturing	20-39
Transport, Electric, Gas	40-49
Wholesale trade	50-51
Retail Trade	52-59
Finance, Insurance and Real Estate	60-67
Services	70-92

For example, 25 represents metal furniture, fixture, 33 (primary metals), 34 (fabricated metal products), 35 (machinery except electrical), 36 (electrical machinery and equipment), 37 (transportation equipment), 38 (precision instruments) and 39 (miscellaneous manufacturing industries) and so on. The machine tool industry has been divided into 9 main SIC categories. These can be further broken down into 76 sub-headings and over 850 specific business types, e.g. 3541 metal cutting, 3542 metal forming, and others.

Sample and data: The first sampling frame was drawn from the British machine tools and equipment directory 1995/96 which consisted of 105 companies. The second sampling frame was drawn from the FAME-CD-ROM database listed under SIC code 3541 (252 companies) and SIC code 3542 (201 companies). FAME is a financial database on CD-ROM containing the information on 270,000 major public and private companies from the Jordan Watch Survey databases. Up to five years of detailed financial information and some descriptive details are also given in the database. FAME contains any company that satisfies one or more of the following criteria:

* Turnover greater than £700,000.
* Shareholders funds greater than £700,000.
* Profits greater than £40,000.

Following comparison between these two directories 82 companies were deleted as they were found in more than one database. A further 42 companies were removed from the database as these companies went into receivership, leaving a net database of 434 companies. In the last week of January 1997, a questionnaire and a personal letter were mailed to the managing director/CEO of each of the 434 companies. Participants were assured of their anonymity; however, if they wished to participate further in the research project, they could do so by placing a tick in the box provided at the end of the questionnaire. Anonymity contributes to unbiased responses. Six weeks after the initial mailing, a second copy of the questionnaire was mailed to those who did not respond. A total of 93 usable questionnaires (73 from first mailing and 20 from second mailing) and 27 unusable responses (e.g. We do not manufacture machine tool, addresses moved, company in receivership) were received at the end of 9 weeks. The overall usable response rate from the first mailing was 18 per cent (73/407) and from the second mailing was six per cent (20/334), leading to a total response rate of 24 per cent. A senior professor of machine tool calibration and the organiser of LAMBDAMAP (1997) felt that the response rate was reasonable. Pitt et al., (1996) from Henley Management College found a response rate of 18 per cent when they surveyed to measure market orientation in the UK.

To assess non-response bias, the last-wave method was used (Filion, 1975, 1976). This method projects the trend in responses across the last two mailing waves. The non-respondents were assumed to respond as did those in the second wave. A series of Chi-square tests indicated no significant differences between first wave respondents and the second wave respondents on any of the measures analyzed (type of industry: CNC, non CNC or both, $\chi 2 = 0.79$, $p > 0.05$; Nature of firms: British or non British firms, $\chi 2 = 0.46$, $p > 0.05$; size of firms: no of employees, $\chi 2 = 1.07$, $p > 0.05$). These results may suggest that the sample is free from non-response bias (Armstrong, 1977). Twelve companies agreed to participate further in the project.

Characteristics of the sample: There are 70 per cent British companies and 30 per cent Joint-venture companies in the sample. The sample consisted of 46 per cent CNC companies, 30 per cent non-CNC and 24 per cent both CNC and non-CNC companies. Companies having turnover less than 10 million, between 10 million and 25 million, and more than 25 million were 65 per cent, 13 per cent, and 20 per cent, respectively. Two percent of respondents did not mention sales turnover of their companies. Companies with employees less than 99, between 100 and 199, and more than 200 represented 62 per cent, 23 per cent, and 15 per cent, respectively. Questionnaires completed by Chief Executive Officers or Managing Directors or Proprietor, Board Level Directors, and senior or middle level managers were 57 per cent, 18 per cent, and 17 per cent, respectively.

Analysis of the Items

Factor analysis, a statistical technique, is used to identify a relatively small number of factors that can be used to represent relationships among a set of several interrelated items. Because the technique relies upon correlations, all variables should be quantitative in nature, though dummy (0-1) variables may be used. More specifically, a factor analysis calculates a series of factors, each of which is a weighted combination of the variables being analysed. KMO (Kaiser- Meyer- Olkin) is an index for comparing the magnitude of the observed correlation coefficients to the magnitudes of the partial correlations. Kaiser (1974) suggests that KMO score above 0.70 is safe to proceed with the factor analysis. In this study, the scores for KMO and Bartlett test of Sphericity are 0.74 and 886.83, respectively.

Principal Component analysis was used to extract factors. In this method, linear combinations of the observed variables are formed. The first principal component is the combination that accounts for the largest amount of variance. The second principal component accounts for the next largest variance and is uncorrelated with the first. Successive components explain progressively smaller portions of the total sample variance, and all are uncorrelated with each other. To achieve a simple structure, often factors are rotated oblique or orthogonally.

The data obtained through the postal questionnaire were subjected to factor analysis to discover underlying dimensions of the market orientation. This was intended to check if these dimensions were consistent with the components of market orientation theory. Table 3.6 shows the factor matrix. Items, which failed to have substantial loading on the factors, e.g. a standardized loading of less than 0.20, were deleted. As expected, two distinct factors were related to customers and competitors, so they were labelled as customer focus (F 1) and competitor focus (F 2). The third distinct factor correlated to the set of items pertaining to responsiveness, consistent with the theory given by Jaworski and Kohli (1993); hence, the name given to this factor was responsiveness (F 3). The fourth and fifth factors are related to customer satisfaction focus (F 4) and market information (F 5), so were named accordingly. Factors six and Factor seven reflected the marketing activities of firms and the negligence of customer service; hence, they were labelled as marketing focus (F 6) and negligence (F 7), respectively.

Table 3.6 **Factor matrix** (coefficients less than 0.20 have been suppressed)

Item	F 1	F 2	F 3	F 4	F 5	F 6	F 7
32	0.78		0.30	0.26			
31	0.77		0.30	0.34			
25	0.71			-0.24		0.26	
35	0.65				0.32		-0.27
41*	0.64	0.32					
18	0.63				0.47		0.32
15	0.58		0.43				
22	0.57		0.23	0.36			
39	0.57	0.36		-0.23	0.45	-0.21	
42	0.52	0.31				0.39	
27	0.52	0.38	0.51				
5		0.80					
14		0.79				0.27	
7	0.24	0.66		0.36			
40	0.37	0.65		-0.36			
37	0.22	0.64	0.30	0.37			
43	0.46	0.52	0.25			0.31	
28*		0.52	0.26				0.33
8		0.51	0.28	0.21		0.29	
10*		0.43		0.35	0.35		
1			0.81	0.22			
11			0.78				
26	0.27		0.67		0.28		
30		0.41	0.60				
36	0.32	0.40	0.58				
12*	0.23		0.48	0.25		0.20	-0.34
24	0.32		0.48	0.40		0.24	
6		0.20		0.80			0.26
16		0.44	0.23	0.63			0.30
3	0.22	0.32	0.38	0.62	0.20	0.21	
20		0.43	0.29	0.54		0.25	
13	0.30	0.39	0.32	0.51		0.29	-0.26
17	0.26				0.74		
4	0.23			0.52	0.68		
9	0.46			0.23	0.66		
38*			0.25	-0.21	0.64		
33*						-0.84	
34*						-0.79	
23	0.22			0.39	0.29	0.58	
29		0.25	0.23		0.42	0.42	0.20
19	0.22					0.31	0.73
21							0.62
2*	0.24		0.22		0.38	0.22	-0.49

* Items finally dropped to achieve higher reliability Cronbach Alpha score.

Except marked items, all items are included for the computation of market orientation composite score.

The set of data produced a seven-factor solution, which accounted for about 67 per cent of the variance. Therefore, all seven factors were included for the development of the market orientation scale. The descriptive statistics of these factors have been reported in Table 3.7.

Table 3.7 Factors statistics for the market orientation scale

Eigenvalue	F 1	F 2	F 3	F 4	F 5	F 6	F 7
Variance	13.14	4.09	2.88	2.52	2.41	1.88	1.80
Mean	30.60	9.50	6.70	5.90	5.60	4.40	4.20
SD	05.43	4.74	4.80	4.69	5.04	4.41	4.87
Alpha	00.89	0.86	0.86	0.86	0.84	0.50	0.61
Std. Alpha	00.89	0.86	0.86	0.87	0.85	0.50	0.61
No of items	11	9	7	5	4	4	3

The Scree plot for factors in Figure 3.3 explains the relationship between Eigenvalue and no of factors.

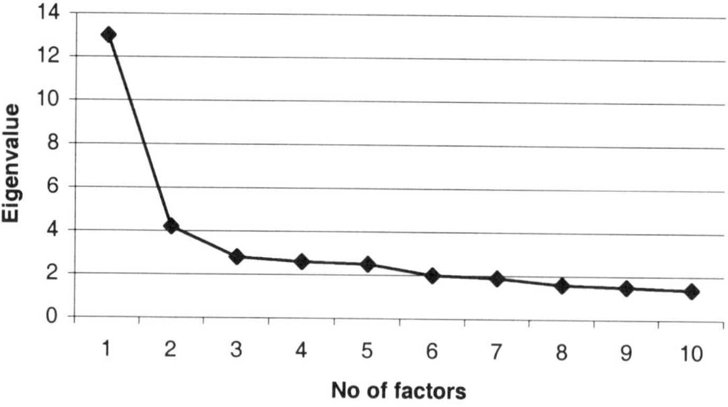

Figure 3.3 Scree plot

Source: The author

Reliability Analysis

In general, there are three methods for the assessment of reliability of a scale: retest, alternative form and internal consistency methods (Peter, 1979). They are explained in the following section.

The retest method is one of the most commonly used methods to estimate the reliability of a scale. In this method, the same test is given to the same subjects after a long period of time and then a correlation is obtained between the scores of the tests. It is presumed that responses to the test will correlate across time because they measure the same variable; however, this method may be contaminated with problems involving respondents' memories. The retest method was not used in the study due to the practical difficulty of scheduling two independent administration of the scale on the same group of respondents.

Alternative form method is similar to the retest method, as it also requires two testing situations with the same subjects; however, it differs from the retest method in that the same test is not given on the second but an alternative form of the test is administered. Because these two forms of the test are intended to measure the same variable, these tests must not differ from each other in any systematic way. The correlation between the alternative forms provides the estimate of the reliability. It is recommended that the two forms of the tests be administered about two weeks apart (Nunnally, 1964). The alternative form method was not used in the study because of the difficulty in constructing two forms of tests, which were systematic in nature and the practical difficulty of administering it in industrial setting.

Internal consistency method refers to the degree to which items in the set are homogeneous. This method is the most popular and practical, as it requires only one scale and one administration of the test. Further, two methods have been widely used to test for the internal consistency of a scale: split-half test and Cronbach Alpha test. The split-half method suffers from a fundamental problem. How scale items should be split into half because depending on how the scale items are split in half, different results will be obtained? Peter (1979) raised an important question as to which reliability coefficient is the real one, as we obtain two reliability scores from the split-half method. Statistical package for social scientists (SPSS) splits the items into two parts by taking the top equal number of items as part one and the remaining equal number of items as part two. In this study, items were entered into SPSS packages in a sequential manner as indicated in Table 3.6. For each factor, split-half reliability test was performed. On the other hand, Cronbach's Alpha test has become almost universally adopted as the Alpha coefficient provides a direct estimate of the mean of all possible split-half tests. Cronbach Alpha suggests how much correlation can be expected between the scale in question and all other possible scales (given the same number of items as the former scale) measuring the same variable. The detailed analysis of reliability for each factor is discussed in the following sections.

Customer Focus Factor (F 1)

Initially, the factor had 11 items. Nunnally (1978) devised a method to evaluate the assignment of items to scales. This approach takes into account the correlation of each item with the scale. Specifically, the item-score to scale-score correlations are used to determine whether an item belongs to the assigned scale. An item is eliminated if it does not correlate highly with the scale. Cronbach Alpha reliability analysis revealed that item 41 (the marketing people in our organization interact frequently with other departments, such as manufacturing, finance, distribution etc.) was least correlated to the rest of the items representing the Factor 1. Hence, item 41 was eliminated to achieve the maximum value of Alpha. To confirm the reliability of the factor, a split-half analysis was performed. Standard Cronbach Alpha and Cronbach Alpha for part 1 and 2 have been reported in Tables 3.8a and 3.8b, respectively. The high Cronbach Alpha scores, ranging from 0.78 to 0.89 and exceeding a threshold of 0.7, as recommended by Cronbach (1975) and Nunnally (1978) suggest that the customer focus factor is reliable. Finally, ten items representing the customer focus factor were retained.

Table 3.8a Reliability analysis for the customer focus factor (F 1)

Item	Corrected item-total correlation	α if item deleted
32	0.79	0.86
31	0.76	0.86
25	0.63	0.87
35	0.64	0.87
42	0.59	0.88
18	0.67	0.87
15	0.58	0.88
22	0.58	0.88
39	0.46	0.89
27	0.62	0.87

α = 0.88, Standard α = 0.89

Table 3.8b Reliability analysis for the customer focus factor (split-half method)

Correlation between forms = 0.78
No of items in part 1 = 5, No of items in part 2 = 5
α for part 1 = 0 .85, α for part 2 = 0.74

Reliability Analysis for the Competitor Focus Factor (F 2)

Cronbach Alpha reliability and split-half reliability tests were performed to check the internal consistency of the competitor focus factor. Initially, the factor had nine items. Individual item-to-total scale correlation suggested that item 28 (Customer complains fall on deaf ears in this business unit) and 10 (we periodically review the likely effect of changes in our business environment, i.e. regulations on customers) should be eliminated for the scale to be more reliable. The overall score for Cronbach Alpha (0.86) did not increase significantly after deleting these two items, which suggested that these items did not contribute significantly to increasing the Cronbach Alpha score. Tables 3.9a and 3.9b report the reliability statistics for the seven-item factor. All values for Cronbach Alpha are above 0.7, recommended to be reliable.

Table 3.9a Reliability analysis for the competitor focus factor (F 2)

Item	Corrected item-total correlation	α if item deleted
5	0.69	0.83
14	0.79	0.82
7	0.74	0.84
40	0.54	0.85
37	0.74	0.83
43	0.59	0.85
8	0.49	0.86

α = 0.86, Standard α = 0.86

Table 3.9b Reliability for the competitor focus factor (split-half method)

Correlation between forms = 0.74
No of items in part 1 = 4, No of items in part 2 = 3
α for part 1 = 0.81, α for part 2 = 0.68

Reliability Analysis for Responsiveness Factor (F 3)

Similar reliability tests performed on the responsiveness factor revealed that item 12 (we often have interdepartmental meetings to discuss market trends and developments) needed removal for the scale to be more reliable. Statistical figures relating to the reliability are shown in Tables 3.10a and 3.10b. Alpha scores above 0.7, as recommended by Nunnally (1978), indicate that the responsiveness factor is reliable. Finally, the scale has six items.

Table 3.10a Reliability analysis for the responsiveness focus factor (F 3)

Item	Corrected item-total correlation	α if item deleted
1	0.60	0.84
11	0.63	0.83
26	0.72	0.82
30	0.64	0.83
36	0.70	0.82
24	0.61	0.84

α = 0.86, Standard α = 0.86

Table 3.10b Reliability analysis for the responsiveness focus factor (split-half method)

Correlation between forms = 0.72
No of items in part 1 = 3, No of items in part 2 = 3
α for part 1 = 0.76, α for part 2 = 0.76

Reliability Analysis for the Customer Satisfaction Factor (F 4)

After conducting both the reliability tests (Cronbach and split-half) on the items it was learnt that it was not possible to increase the reliability of the factor on the basis of deleting any item. The scale has five items. Therefore, it may be suggested that each of these items contributed significantly to the reliability of the customer satisfaction factor. The reliability statistics are presented in Tables 3.11a and 3.11b. Values for Alphas in excess of 0.7 suggest the reliability of the factor.

Table 3.11a Reliability analysis for the customer satisfaction focus factor (F 4)

Item	Corrected item-total correlation	α if item deleted
6	0.64	0.85
16	0.64	0.85
3	0.76	0.82
20	0.71	0.83
3	0.73	0.83

α = 0.86, Standard α = 0.87

Table 3.11b Reliability analysis for the customer satisfaction focus factor (split-half method)

Correlation between forms = 0.73
No of items in part 1 = 3, No of items in part 2 = 2
α for part 1 = 0.79, α for part 2 = 0.80

Reliability Analysis for the Market Information Focus Factor (F 5)

These reliability tests suggested that item 38 (in our organization customer calls are disseminated across other departments) should be eliminated to achieve a higher reliability score. Tables 3.12a and 3.12b suggest that the market information focus factor is reliable, as evidenced by high Cronbach Alpha scores. The fact that there are only three items on the market information factor and that the scale is statistically reliable suggests these three items are good representative of the factor. Further, there are high correlations among all the items. Although variance explained by the factor is low, it is worth analysing in an industrial setting.

Table 3.12a Reliability analysis for the market information focus factor (F 5)

Item	Corrected item-total correlation	α if item deleted
17	0.70	0.80
4	0.70	0.80
9	0.75	0.76

α = 0.84, Standard α = 0.85

Table 3.12b Reliability analysis for the market information focus factor (split-half method)

Correlation between forms = 0.75
No of items in part 1 = 2, No of items in part 2 = 1
α for part 1 = 0.76, α for part 2 = not applicable

Reliability Analysis for the Marketing Focus Factor (F 6)

The factor failed to meet the reliability criteria established by Nunnally (1978). This factor initially had four items. The correlation matrix revealed that item 33 (in our company, there is a little distinction between sales and marketing) and item 34 (in our company, marketing's most important job is to promote our product and services to our customers) were negatively related to the rest of the items, contributing to lowering Cronbach Alpha score. To increase the score, these two items were eliminated, leaving only two items in the scale. Although the factor has a Cronbach Alpha score below 0.7 as recommended by Nunnally (1978) for a scale to be reliable, this factor was taken into account for the explanation of variance in the market orientation scale. Eigenvalue for the factor is more than one and the factor explained more than four per cent of the variance in the market orientation scale. This factor is unique in that respondents in the machine tool industry separate market orientation from sales orientation. It appears that more items should be included to the scale for the factor to be statistically reliable. Statistical figures are shown in Table 3.13.

Table 3.13 Reliability analysis for the marketing focus factor (F 6)

Item	Corrected item-total correlation	α if item deleted
23	0.33	not applicable
29	0.33	not applicable

α = 0.50, Standard α = 0.50

The reliability analysis for the marketing focus factor via split-half method is not applicable

Reliability Analysis for the Negligence Factor (F 7)

The negligence factor failed to meet the criterion laid down by Nunnally (1978). Initially, the factor had three items. The correlation matrix suggested that item two (individuals from our manufacturing department interact directly with customers to learn how to serve them better) correlated negatively with rest of the items. Hence, item two was eliminated to improve the reliability score. The score could not be improved further due to lack of items on the factor. Although the Cronbach Alpha scores for the marketing focus and negligence factors are below 0.7 as suggested by Cronbach (1975), they are retained for the calculation of preliminary market orientation score. The reliability figures are presented in the Table 3.14.

Table 3.14 Reliability analysis for the negligence factor (F 7)

Item	Corrected item-total correlation	α if item deleted
19	0.44	not applicable
21	0.44	not applicable

$\alpha = 0.61$, Standard $\alpha = 0.61$

The reliability analysis for the negligence factor via split-half method is not applicable

Validity Assessment

Validity is defined as the extent to which a scale measures what it is intended to measure. Validation of a scale is particularly important because it is quite possible for a scale to be relatively valid for measuring one kind of variable but entirely invalid for assessing another variable. Thus, it is not only sufficient to validate the measuring instrument independently but also it is necessary to validate it in the context for which it is being used. There are generally three kinds of validity checks: content, construct, and criterion validity. They are described below.

Content Validity

Fundamentally, content validity depends on the extent to which a measurement reflects a specific domain of contents. A scale can be said to have content validity if there is a general agreement among researchers that the items cover almost all aspects of a variable being measured. In the study, the market orientation scale may be assumed to have content validity because of the following reasons:

* As a result of the comparative analysis, items were taken from three different market orientation scales, which are supposed to tap the domain of the scale.
* These three market orientation scales were the results of exhaustive literature reviews and extensive interviews with field managers and academicians. Further, these scales were tested and found reliable and valid.
* The judgement of content validity is subjective; however, the procedure used in the study may ensure content validity of the scale. Indeed, Bohrnstedt (1970) argues that while we enthusiastically endorse the procedure, we reject the concept of content validity on the grounds that there is no rigorous way to assess it.

Construct Validity

Construct validity is concerned with the extent to which a scale relates to other scales designed to measure the same theoretical construct. There are two methods to assess the construct validity of a scale. They are as follows.

Convergent validity: To meet the convergent and discriminant validity, we would expect, in the factor analysis, that all items representing a concept should load strongly on one factor to satisfy the requirement of the convergent validity and weakly on all other factors to satisfy the requirements of the discriminant validity (Balkrishnan 1996). By using this approach, the market orientation scale was refined by eliminating items that either did not load strongly on any factor, or loaded on more than one factor. Table 3.15 suggests that the market orientation scale meets both convergent and discriminant validity.

Discriminant validity: Further, to test for the discriminant validity, the questionnaire incorporated a brief description of two companies about their set of activities. Respondents were asked to indicate the extent to which their business units resembled these companies by distributing 100 points between them. Thus, if the business unit was like Company A and only remotely like Company B, one could allocate 90 points to Company A and 10 points to Company B. The discriminant validity question was:

* Company A relies on its salespeople to use a variety of selling techniques for getting customers to say yes. The primary emphasis in the company is on selling. Customer satisfaction is considered important but the emphasis is on going out and pushing the company's products.
* Company B does a lot of research to learn the concerns of its customers and responds by developing new products and marketing programs. The emphasis is on understanding why customers act and feel the way they do and exploiting this knowledge. Selling is considered important, but the emphasis is on making products that will almost sell themselves.

Company A:_____ points Company B:_____ points. (Total = 100)

Source: Jaworski and Kohli (1993)

A correlation analysis was conducted between the scores relating to Company A, Company B and the market orientation factors. Table 3.15 represents the results of the correlation analysis. It suggests that the market orientation scale is significantly and positively related to Company B and significantly and negatively correlated to Company A. Further, all seven factors of market orientation are positively related to Company B and negatively correlated to Company A. This pattern of correlations suggests that the market orientation scale discriminates between Company A and Company B. Company A and B represent sales- and market-oriented companies, respectively.

Table 3.15 Discriminant validity analysis

Factors	Company A	Company B
F 1	-0.35**	0.31**
F 2	-0.25*	0.32**
F 3	-0.11	0.10
F 4	-0.05	0.13
F 5	-0.15	0.10
F 6	-0.04	0.03
F 7	-0.09	0.09
MO Scale	-0.20	0.20*
Company A	1.00	-0.91**
Company B	-0.93*	1.00

** $p < 0.01$, * $p < 0.05$

Criterion Validity

Criterion-related validity is concerned with the extent to which the score on a scale is related to an independent scale of the relative criteria. This is also referred as external validity. The criterion-related validity was evaluated by examining multiple regression correlation coefficients between the scores on the market orientation scale and an independent measure, which asked respondents the extent to which they perceived their companies were market oriented on a 7-point Likert scale, 1 being not market-oriented and 7 being market-oriented. As evident from Table 3.16, a significant correlation (0.65, $p < 0.00$) suggests that the scale has a criterion-related validity.

Table 3.16 Dependent variable: overall market orientation

Independent Variable: market orientation
Multiple R = 0.65, R^2 = 0.43, Adj. R^2 = 0.42, Standard error = 0.65
ANOVA, F= 69.79, $p < 0.00$

Variable in the equation	B	SE	β	t	significance
Market orientation	0.48	0.05	0.65	8.3	0.00
(Constant)	2.45	0.29		8.2	0.00

Table 3.17a Reliability analysis for the market orientation scale

Item	Corrected item-total correlation	α if item deleted
32	0.58	0.93
31	0.59	0.93
25	0.51	0.93
35	0.47	0.93
18	0.49	0.93
15	0.47	0.93
22	0.53	0.93
39	0.43	0.93
42	0.56	0.94
27	0.61	0.93
5	0.50	0.93
14	0.55	0.93
7	0.61	0.93
40	0.46	0.93
37	0.77	0.93
40	0.46	0.93
37	0.77	0.93
43	0.64	0.93
8	0.39	0.93
1	0.50	0.93
11	0.39	0.39
26	0.61	0.93
30	0.51	0.93
36	0.68	0.93
24	0.60	0.93
6	0.45	0.93
16	0.58	0.93
3	0.75	0.93
20	0.67	0.93
13	0.71	0.93
17	0.50	0.93
4	0.62	0.93
23	0.38	0.93
29	0.51	0.93
19	0.39	0.93
21	0.41	0.93

α = 0.93, Standard α = 0.94

Market Orientation Scale

The factor analysis produced seven distinct factors. These factors were labelled as:

* Customer Focus
* Competitor Focus
* Responsiveness
* Customer Satisfaction Focus
* Market Information
* Marketing Focus, and
* Negligence

To check the reliability of the market orientation scale, all items pertaining to individual factor have been combined to form a composite market orientation scale, which is consistent with the methodology adopted by several researchers (Ruekert, 1992; Cronin and Taylor, 1992; Atuahene-Gima, 1995; Greenley, 1995; Gray et al. 1998; Singh, 2001). The market orientation scale has 35 items. The Cronbach Alpha and standard Cronbach Alpha scores for the market orientation scale are 0.93 and 0.94, respectively, suggesting the scale to be reliable. This reliability is further confirmed by performing split-half reliability test on the scale. Alpha scores for part 1 and part 2 are 0.89 and 0.89, respectively, indicating reliability of the scale. Tables 3.17a and 3.17b show the values for the Alpha.

Table 3.17b Reliability analysis for the market orientation scale (split-half method)

Correlation between forms = 0.78
No if items in part 1 = 18, No of items in part 2 = 17
α for Part 1 = 0.89, α for Part 2 = 0.89

Concluding Remarks and Limitations

The objective of the chapter was to redesign a reliable and valid scale for measuring market orientation by combining the items from the most relevant scales proposed in literature in the last decade. The new scale is intended to be representative of comprehensive set of market-oriented activities in the context of machine tool industry.

This chapter described a seven-factor market orientation scale. These factors were labelled as customer focus, competitor focus, responsiveness, customer satisfaction focus, market information focus, marketing focus and negligence, and have 10, 7, 6, 5, 3, 2 and 2 items, respectively. Total variance explained by these factors is 67 per cent. The last two factors, i.e. 'marketing focus' and 'negligence' have only two items each, which accounted for, taken together, for more than eight per cent of the variance in the scale. Therefore, these two factors were retained at this stage for the computation of the composite market orientation score. Coefficient Alpha methodology was used to assess the reliability of the scale and its factors. Further, the market orientation scale was found to have content, convergent, discriminant and criterion-related validity.

In the next chapter, a parsimonious market orientation scale is developed. Next, the effect of market orientation and its factors is assessed on various performance indicators, such as return on investment, sales growth, market share, new product success, customer retention and global presence. Further, there is an examination as to whether the market orientation-performance relationship is moderated by external factors, such as market turbulence, competitive intensity and technological turbulence.

Chapter 4

Statistical Data Analysis

Introduction

The overall aim of the chapter is to examine if there is a relationship between market orientation and business performance - measured by return on investment, sales growth, market share growth, new product success rate, customer retention rate and global market presence of companies - and if the relationship is moderated by market turbulence, technological turbulence and competitive intensity. Statistical techniques, such as subgroup analysis, moderated regression and analysis of variance have been employed to detect the effects from these moderators. Next, a parsimonious model representing the market orientation has been developed. It was discovered that the market orientation was best represented by the factors relating to customer, competitor, satisfaction and marketing orientations. Finally, the effect of each individual factor has also been examined on the overall performance of companies.

The Variables

The Independent Variables

Competitor concentration: This refers to an assessment of the degree of monopolistic power within the marketplace (Porter, 1985; Houston, 1986). Economic theory suggests that a high concentration should lead to better performance for the major competitors as they recognize the benefits of avoiding price competition (Slater and Narver, 1994). This is a single item scale.

This item was measured on a Likert scale. Marketing Researchers' adaptation of the method of summated ratings developed by Likert (1967) is an extremely popular means for measuring attitudes because it is simple to administer. With the scale, respondents indicate their attitude by checking how strongly they agree or disagree with carefully constructed statements ranging from very positive to very negative attitudes toward some statements. For the study, respondents were asked to indicate the extent to which they agreed with the question - our competitors are relatively weak - on a seven-point Likert scale, one being strongly disagree and seven being strongly agree.

Customer power: This was measured as the ability of customers to obtain prices from the suppliers that are lower than the suppliers wish to charge (Narver and Slater, 1990). A strong customer power indicates lower profits for suppliers; therefore, a negative relationship is expected with the performance. It is a single item scale - customers have power to negotiate lower prices from seller - which was measured on a seven-point Likert scale.

Relative size: This is the company size, measured as sales volume relative to the largest competitor (Porter, 1980). A positive relationship is expected between a firm's size and its business performance. Respondents were asked to indicate the size of their businesses relative to their largest competitor on a seven-point Stapel scale, where -3 being much smaller, 0 being the same as their competitors and +3 being much bigger.

The Stapel scale originally was developed in the 1950s to measure simultaneously the direction and intensity of an attitude. Modern versions of the scale use a single adjective as a substitute for the semantic differential when it is difficult to create pairs of bipolar adjectives. The modified Stapel scale places a single adjective in the centre of an even number of numerical values, ranging, usually, +3 to -3 (Crespi, 1961). It measure how close to or distant from the adjective a given stimulus is perceived to be. For accurate measurement of the variables, few control variables and all performance indicators were measured on Stapel scale. Scales associated with variables are indicated in the following sections.

Relative cost: This is a measure of per unit operating costs, compared to the largest competitor within its market segment (Porter, 1980). A high cost denotes a cost disadvantage; hence, a negative association is expected between relative cost and business performance. Respondents were asked to indicate the average total operating cost of their businesses relative to their largest competitor in their main market segment on a seven-point Stapel scale, where -3 being much smaller, 0 being same as competitor and 3 being much bigger.

Ease of market entry: This refers to the extent to which there is a possibility that new market entrants would earn satisfactory profits in their principal market segments in the first three years (Scherer, 1980). Easy entry indicates a disadvantage to current competitors (Porter, 1980); thus, a negative link is expected with performance. Respondents were asked to indicate on a seven-point scale, 1 being very easy, 7 being very difficult, how easy it was for a new entrant to earn satisfactory profits in their main market segment within three years after entry.

Dominance of sales: This refers to the extent to which respondents' markets are dominated by their competitions. Respondents were asked to indicate their choices on a seven-point Likert scale, one being sales are dominated by less than four companies, four being sales dominated by about ten companies and seven being sales dominated by more than ten companies. If there were many players in the same segment, a negative sign would be expected with performance.

Market orientation: The market orientation was measured on the 35-item scale, which was represented by the seven factors: customer, competitor, responsiveness, satisfaction, market information, marketing orientation and negligence. Consistent with Lusch and Laczniak (1987), the selected items tapped the comprehensive activities relating to the needs and preferences of customers and end-users. Each item was scored on a seven-point scale, one being strongly disagree and seven being strongly agree.

The Dependent Variables

The five indicators were considered proxy for the business performance: Return on Investment (p-roi), Sales Growth (p-sg), Market Share (p-mktshr), New Product Success (p-nps), Customer Retention (p-cusret) and Global Presence (p-global). These five variables were measured on a Stapel, -3 being much smaller, zero being the same as their competitors and +3 being much bigger. Data collection on these dependent variables were based on the managing directors' subjective response to questions that assessed whether results of return on investment, sales growth, market share growth, new product success rate, customer retention rate and global presence were above or below their expectations over the last three years. At the pre-test stage of questionnaires, it was learnt that respondents were more likely to provide a more accurate estimate of their companies' performance over a period of three years, as it takes at least three years to realise the effect of any changes made to the machines on the performance of companies. The last item, global presence refers to the extent to which a company is successful in making its products or services present in foreign markets. A single item - our markets are becoming increasingly global - was used in the questionnaire to measure the global presence variable.

Subjective Versus Objective Approach

For the data collection, a subjective approach was employed due to the difficulty in obtaining objective data from documentary sources and the reluctance of firms to divulge information, which was classified as confidential. Researchers who adopted both concepts reported a strong association between objective measures and subjective responses (Venkatraman and Ramanujan, 1986; Robinson and Pearce, 1988). Jaworski and Kohli (1993) utilized both methods and obtained reliable responses for their variables. The argument is that this type of subjective performance measure, with an anchor relative to expectations, allows for greater comparability across types of industry and situations with varying standards of satisfactory performance (Pelham and Wilson, 1996). Further, interpreting objective measures for companies can be rather complicated. For instance, low profits or operating losses in small, growth-oriented businesses may not be a sign of poor management if this apparent unsatisfactory performance is due to massive investments in a product or market development (Covin and Slevin, 1989).

Single Versus Multiple Respondent Technique

Further, single-respondent methodology was preferred for collection of data because of three reasons: first, given the small size of the companies, it was possible to achieve a more accurate view of an organization's overall level of market orientation from a senior manager, particularly because market orientation is instituted at corporate level; second, by not distributing questionnaires at different management levels in an organization, it avoids over representation of certain department, a it practically difficult to know which company has which departments; and third, a single questionnaire is relatively easy to administer in that if a corporate executive is busy or not available, they can choose a proxy respondent for them. In any methodology whether a single-respondent or multi-respondent, obtaining a representative sample was important for capturing the overriding market-oriented culture of organizations. The majority of studies are based on single-respondent sample. Kohli et al. (1993) employed both single- and multiple-respondent samples in their study. Conduit and Mavondo (2001) used only multi-respondent techniques for collection of data from a very small set of samples.

Analysis of the Dependent Variables

Before conducting the regression analysis, it is important to analyze the nature of the dependent variables. There are two basic assumptions associated with dependent variables: first, a dependent variable should not have outliers. An outlier is a value of a dependent variable, which in some way is inconsistent with the other sample values of the dependent variable. Hence, inclusion of outliers can affect the estimates and prediction of regression models. And second, a dependent variable should ideally be normally distributed which means that it should be symmetrical about the mean. If dependent variables do not follow a normal distribution curve, this may result in excessively large error terms for certain cases, reducing the reliability of the regression model. The requirement of a normally distributed dependent variable is not as essential as the removal of outliers from the model. Norusis (1993) suggests that because it is almost impossible to find data that are exactly normally distributed, which is the requirement for most statistical tests, it is sufficient that data are approximately normally distributed. In the following sections, the characteristics of individual dependent variables are analyzed to establish whether they violated either of these assumptions, i.e. the pattern of values should be approximate to normal distribution and there should be no outliers.

The analysis of the dependent variable revealed that the distribution of values for p-roi nonsignificantly negatively skewed ($t = 6.89$, $p < 0.00$; and skewness = -0.58). A Kolmogorov-Smirnov (K-S) test was conducted to test whether the p-roi could be assumed as being normally distributed. The null hypothesis is that there is no significant difference between the sample cumulative distribution function and the hypothetical cumulative distribution function. The results of this test (K-S = 0.10, $p < 0.00$) led the null hypothesis to be rejected. A boxplot was produced to establish whether any outlier could be identified in the distribution of values for the p-roi. Box plot (not shown) revealed that there were a couple of outliers. To make the data follow a normally distributed curve, outliers were removed from the data. Characteristics of the p-roi were analyzed again following the removal of these outliers.

The removal of these outliers reduced the skewness from -0.58 to -0.23, which also reflected the reduced gap between the values of the mean and median of the data without outliers. Although the results of K-S test (K-S = 0.13; $p < .00$) revealed that the variable is still not exactly normal ($t = 5.12$, $p < 0.00$) but did follow the normal distribution function. Norusis (1993) suggests that F tests used in the regression analysis are usually insensitive to moderate departure from normality.

Similarly, other dependent variables were analyzed to examine if they violated either of the assumptions for their suitability to be included in the regression models. Histograms and box plots were constructed for the rest of the dependent variables, i.e. p-sg, p-mktshr, p-nps, p-cusret, p-global. The box plots (not shown) for these variables suggested that the scores less than 4, 3, 3, 4 and 2 required removal from p-sg, p-mktshr, p-nps, p-cusret, and p-global, respectively to transform the data more normally distributed. By removing these outliers from these dependent variables, the skewness of these variables was drastically reduced, as evidenced by a little difference between the mean and median for each dependent variable. A K-S test was performed on each dependent variable to ascertain the normality of the data, which suggested that the data followed a normal distribution function with some degree of acceptable departure from normality. Based on the t-test statistics for the dependent variables, the departure of the data from normality is within the acceptable range for the construction of regression model.

The overall performance variable is a composite score of the five dependent variables mentioned in the previous section. Computation of the score consistent with previous studies (Ruekert, 1992; Gary et al., 1998; Singh, 2001; Singh, 2004) The K-S statistics for the overall performance variable (K-S = 0.07, $p < 0.19$) indicates that the variable is more normally distributed than the rest of the dependent variables taken separately.

Regression Analysis

A linear regression analysis investigates a straight-line relationship of the type $Y = C + \beta * X$, where Y is the dependent variable, X is the independent variable, and C and β are two constants to be estimated. The symbol C represents the Y intercept and β is the slope coefficients. The slope β is the change in Y due to a corresponding change in one unit of X. Further, a multiple regression analysis is used that allows for simultaneous investigation of the effects of two or more independent variables on a dependent variable. The aim is to find the best means for fitting a straight line to the data. The *least-square method* is a relatively simple mathematical technique that ensures that the straight line most closely represent the relationship between dependent and independent variables. The coefficient of determination, R^2, measures that part of the total variance of dependent variable (Y) that is accounted for by independent variables (X_1, X_2 ...) in the equation. There are different methods to the multiple regression analysis, which differ from one another in a manner in which independent variables are entered into the regression equation. A brief description of few methods is presented in the following section.

Forward Selection Method

In this method, variables are entered in the model one at a time based on entry criteria. In the beginning, there is no independent variable in the model. The first variable considered for entry into the equation is the one, which has the largest positive or negative correlation with the dependent variable and is above the significant threshold. The significance of each independent variable is then calculated in the model and the least significant variable is removed from the model. The process is iterated till all the significant independent variables have been included in the regression model. The procedure stops when there are no other variables that meet the entry criterion. However, the forward selection method has a limitation. Because the variables are entered and removed on the basis of the certain criteria (usually F-to-enter (FIN) = 3.84 or probability-to-enter (PIN) = 0.05), the significance of variables already in the model may change with the addition of another variable in the equation. Therefore, a variable, which was entered into the equation because of its significance, may be removed after the inclusion of an additional variable. This may be caused by the correlations between the independent variables. As a result of the overall correlation among independent variables, some significant variables may not be in the final equation.

Backward Elimination Method

While forward selection begins with no independent variables in the equation and subsequently enters them, backward elimination starts with all variables in the equation and subsequently removes them. Instead of entry criteria, removal criteria are used. The significance of each independent variable is calculated in the model and the least significant variable is removed from the model. The first criterion is the minimum F value that a variable must have in order to remain in the equation. Variables with F values less than F-to-remove (FOUT) are eligible for removal. The second criterion is the maximum probability of F-to-remove (POUT) that a variable can have. The default values for FOUT and POUT are 2.71 and 0.10, respectively. The default criterion is the probability of F-to-remove. The process is repeated until all nonsignificant variables have been removed from the model. The objective of the model is to include only those variables, which meet the predefined criteria. However, the method has the same limitation as forward selection in that the variables are removed from the model, the significance of those variables, which have already been excluded, may change. Hence, the method may generate a model excluding significant variables.

Stepwise Regression Method

The stepwise regression method is the combination of forward and backward procedures. The first variable is selected in the same manner as in forward selection method. If the variable fails to meet the entry requirement (either FIN or PIN), the procedure terminates with no independent variables in the equation. If it passes the criterion, the second variable is selected based on the highest partial correlation and it enters into the equation. The difference between the stepwise method and the forward method is the way independent variables are included in the model. In forward selection method, once a variable has been included in the model, it cannot be removed. However, in the stepwise regression, a variable is excluded from the model if it becomes nonsignificant following the addition of other variables to the equation. Variables are removed when they meet the removal criterion. To prevent the same variable from being repeatedly entered and removed, POUT is kept higher than PIN. The stepwise method, which is a combination of both forward and backward approaches, ensures that all significant variables are added, and that all the nonsignificant variables are excluded. This technique also ensures that any correlation between independent variables is accounted for in the regression model.

Enter Method

In this method, all variables are entered into the regression equation irrespective of the significance of their contribution to the overall model. Problems are encountered in the model when there are too many variables. On the other hand, the method is useful when it is intended to examine the significance of each variable in the presence of the rest of the independent variables. This could be a desirable technique in a situation where it is difficult to remove a variable from an equation unless it has met certain criteria.

Because the aim of the study is to find the partial correlation coefficients between each independent variable and dependent variable, the Enter method was used; however, it is recognized that all the procedures do not always result in the same equation.

Tests for Main Effects

To test the hypotheses (H1a through H1f), the following multiple regression equations were estimated:

$$Y_1 = \text{Constant} + \beta_1X_1 + \beta_2X_2 + \beta_3X_3 ... + \beta_{10}X_{10}$$
$$Y_2 = \text{Constant} + \beta_1X_1 + \beta_2X_2 + \beta_3X_3 ... + \beta_{10}X_{10}$$
$$Y_3 = \text{Constant} + \beta_1X_1 + \beta_2X_2 + \beta_3X_3 ... + \beta_{10}X_{10}$$
$$Y_4 = \text{Constant} + \beta_1X_1 + \beta_2X_2 + \beta_3X_3 ... + \beta_{10}X_{10}$$
$$Y_5 = \text{Constant} + \beta_1X_1 + \beta_2X_2 + \beta_3X_3 ... + \beta_{10}X_{10}$$
$$Y_6 = \text{Constant} + \beta_1X_1 + \beta_2X_2 + \beta_3X_3 ... + \beta_{10}X_{10}$$
$$Y_7 = \text{Constant} + \beta_1X_1 + \beta_2X_2 + \beta_3X_3 ... + \beta_{10}X_{10}$$

Where, Y_1=Return on Investment (p-roi), Y_2=Sales Growth (p-sg), Y_3=Market Share growth (p-mktshr), Y_4=New product Success (p-nps) rate, Y_5=Customer Retention (p-cusret), Y_6=Global Presence (p-global), Y_7=Overall Performance (p-ov); and, X_1 through X_{10} correspond to (1) Market Orientation (2) Market Turbulence (3) Competitive Intensity (4) Competitors Concentration (5) Technological Turbulence

(6) Customer Power (7) Relative Size (8) Relative Cost (9) Ease of Entry and (10) Dominance of Sales, respectively.

Table 4.1 reports the means and standard deviations of the independent variables included in the equation.

Table 4.1 Descriptive statistics

Variables	Mean	Standard Deviation
MO-Scale	5.02	1.41
Market Turbulence	4.43	1.47
Competitive Intensity	4.53	1.54
Competitors Concentration	3.15	1.57
Technological Turbulence	5.45	1.15
Customer Power	5.63	1.04
Relative Cost	3.13	1.49
Ease of Entry	5.51	1.26
Dominance of Sales	4.09	1.82
Sales Growth	5.71	1.10
Market Share	5.43	1.13
New Product Success Rate	5.29	1.04
Return on Investment	5.00	1.40
Overall Performance	5.47	1.19
Customer Retention	6.04	1.01
Global Presence	5.57	1.57

Multicollinearity

One of the difficulties using the multiple regression analysis is the existence of multicollinearity among independent variables. This condition refers to a situation in which some of the independent variables are highly correlated with each other. In such a situation, collinear variables do not provide any new information; therefore, it becomes difficult to separate the effect of such variables on dependent variables. As a result, the values of the regression coefficients for the correlated variable may be influenced drastically, depending upon which variables are included in the model. One method of measuring collinearity is the calculation of variance inflation factor (VIF) for each variable. The VIF was calculated by the following equation:

$$VIF_i = \frac{1}{(1 - R_i^2)}$$

Where, R^2_i represents the coefficient of determination of independent variable X_i with other X variables.

If a set of independent variables is uncorrelated, the score for the variance inflation factor (VIF) should be equal to one. Marquandt (1980) suggests that if the value for the VIF is greater than 10, there may be a significant correlation between variable X_i and other independent variables. Snee (1973) recommends that an alternative method to the least-square regression should be used if the score for the VIF exceeds five. One of the ways to keep the VIF below five is the detection of outliers. These outliers give rise to inflated VIF. In this study, the multiple regression analysis was used after having examined the scores for the VIFs. It was revealed that the values for the VIF were below five; therefore, it was safe to proceed with the regression analysis. Scores for the VIF are reported in Table 4.2.

Main Effect Results and Discussions

Table 4.2 Main effect results (standard β coefficients and standard errors)

	Sales Growth	Market Share Growth	NPS	ROI	Overall Perf.	Cust. Retn.	Global Pres.
MO	0.24^{*}	0.25^{***}	0.32^{***}	0.36^{***}	0.16	0.20^{*}	0.31^{***}
Scale	(0.07)	(0.09)	(0.09)	(0.12)	(0.06)	(0.06)	(0.15)
Market	-0.21^{*}	0.00	0.27^{***}	0.03	0.09	-0.01	-0.16
Turbulence	(0.06)	(0.08)	(0.08)	(0.10)	(0.06)	(0.05)	(0.12)
Competition	-0.37^{***}	-0.25^{**}	-0.22^{**}	-0.10	-0.25^{**}	-0.45^{***}	-0.16
Intensity	(0.05)	(0.06)	(0.06)	(0.09)	(0.05)	(0.04)	(0.10)
Competition	-0.05	-0.23^{***}	-0.08	-0.18^{*}	0.09	-0.07	-0.07
Concentration	(0.03)	(0.04)	(0.04)	(0.06)	(0.03)	(0.03)	(0.07)
Technological	-0.11	-0.02	-0.12	-0.29^{***}	-0.24^{**}	-0.27^{**}	0.09
Turbulence	(0.05)	(0.07)	(0.07)	(0.10)	(0.05)	(0.05)	(0.12)
Customer	0.14	-0.18^{**}	0.00	-0.00	-0.03	-0.01	-0.03
Power	(0.06)	(0.08)	(0.08)	(0.11)	(0.06)	(0.05)	(0.13)
Relative	0.25	0.24^{*}	0.29^{**}	0.39^{***}	0.22^{**}	0.19	-0.01
Size	(0.03)	(0.04)	(0.04)	(0.06)	(0.03)	(0.03)	(0.07)
Relative	-0.14	-0.26	-0.22^{*}	-0.36^{***}	-0.29^{***}	-0.05	-0.35^{**}
Cost	(0.05)	(0.06)	(0.06)	(0.09)	(0.04)	(0.04)	(0.10)
Ease of	-0.10	-0.02	-0.05	0.16	-0.11	-0.08	0.02
Entry	(0.05)	(0.06)	(0.06)	(0.09)	(0.05)	(0.04)	(0.11)
Dominance	-0.13	0.09	-0.13	0.04	0.06	-0.12	-0.19^{*}
Sales	(0.03)	(0.04)	(0.04)	(0.05)	(0.03)	(0.03)	(0.07)
F	6.24^{***}	5.42^{***}	8.58^{***}	8.14^{***}	14.62^{***}	10.08^{***}	5.78^{***}
R^2	0.48	0.43	0.53	0.52	0.65	0.59	0.43
Adj. R^2	0.40	0.35	0.47	0.45	0.61	0.53	0.36
VIF_{Max}	1.71	1.69	1.88	1.86	1.75	1.99	1.74

$^{*}p < 0.05$, $^{**}p < 0.01$, $^{***}p < 0.00$

Hypotheses H1a through H1f examined the impact of market orientation on the dimensions of business performance. Table 4.2 contains the result of the tests for the main effect of market orientation on business performance.

Market orientation is the only significant predictor for all the dependent variables except overall performance. Although market orientation is not a significant predictor for the overall performance, it is positively correlated to the overall performance. The regression coefficient ($\beta = 0.16$) between market orientation and overall performance provides a mixed support for the hypothesis H1 that the market orientation leads to better business performance. The hypothesis is accepted. This finding is consistent with the study of Kohli and Jaworski (1993), Caruana (1995) and Pitt et al. (1996), which suggest that a degree of market orientation enables an organization to perform better than its counterparts.

Further, market orientation has a significant ($p < 0.00$) and positive ($\beta = 0.36$) effect on return on investment of machine tool companies; therefore, the hypothesis H1a that market orientation is positively related to return on investment is accepted. This result is consistent with the studies of Narver and Slater (1990), Slater and Narver (1994) and Pelham and Wilson (1996) but inconsistent with Greenley's (1995), suggesting that the machine tool manufacturing firms with higher degree of market orientation are likely to be more profitable. Because a market–oriented firm should have lower expenses associated with new product failures and correcting quality control problems, it should reflect high return on investments. This is particularly true in case of small industrial firms, which are noted for their low level of formal planning (Sexton and VanAuken, 1985) and for formal market research (McDaniel and Parasuraman, 1986). It may be noted that in some cases, firms may not be able to realize profit from sales due to high competition (for example, cost cutting strategy pursued by larger firms), but may generate a profit by providing specialised services to those customers at extra cost. Further, it appears that executives have innovative ideas to open small shops on the premises of their customers who ordered significant machine tools and repair services on a regular basis. It was economical for these executives to have the concept of 'shop-inside-workshop' to be market-oriented and, therefore, maximize return on investment (Singh, 2003).

Hypothesis H1b that market orientation is positively related to sales growth is supported because the market orientation has a significant ($p < 0.05$) and positive ($\beta = 0.24$) effect on sales growth. While this result is consistent with the study of Slater and Narver (1994), it is inconsistent with the findings of Greenley (1995) and Pelham and Wilson (1996). Greenley argues that market orientation has no influence on sales growth when selling power lies with the company, because in this situation companies do not feel the need for being market oriented. On the other hand, Pelham and Wilson (1996) discovered that the impact on market orientation on sales growth and market share might be indirect through the success of new products. In the context of the machine tool sector, it is felt that a market orientation should lead to higher sales growth on the ground that industrial buyers rely substantially on the performance of the previously purchased machines. Good performance of a machine generates customer satisfaction, loyalty and a good image for the company. These factors have been found to have links with customer retention, which in turn has a positive impact on repeat purchases, leading to higher sales growth. Sales growth via repeat purchase is particularly important for machine tool manufacturers because they rely significantly on sectors, such as the automotive industry, aviation, processing plants and others.

Certainly, a reduction in demand from these sectors would have a negative impact on the sales growth of these companies.

Market orientation has a significant ($p < 0.00$) and positive ($\beta = 0.25$) impact on market share, thus the hypothesis H1c that the market orientation is positively related to market share is supported. This result is inconsistent with the study of Kohli and Jaworski (1993), which suggests that few high performing companies may deliberately pursue a focus strategy and may be unconcerned about their market share positions. Further, there may be a lagged effect of market orientation on market share, which might not have been detected by the cross-sectional design of their study. In the machine tool industry, the market share of companies is usually constant over a long period of time as capital items are less frequently purchased than consumer items. However, market share can be increased by the quality and the superior performance of machines. Capital items like machine tools are generally purchased on two occasions. First, when there is a replacement for the old machine, and second when there is an expansion of the manufacturing firms to increase production capacity. So the durability of the machines may play an important role in the reordering of new machines, which in turn may increase their market share. Durability indicates two aspects of a machine: first, higher durability means higher efficiency, realized through the long life of the machine which over a period of time will have positive impact on financial business performance; and second, high durability of machines will avoid he need for purchase of new machines. Therefore, if a company has a reputation for its high durable machines, it is relatively easier to increase market share because it is likelihood that potential customers will seek to purchase new machine from the suppliers.

Market orientation has a significant ($p < 0.00$) and positive ($\beta = 0.32$) impact upon new product success, thus the hypothesis H1d that the market orientation is positively related to new product success is supported. This result is consistent with the findings of Pelham and Wilson (1996), Slater and Narver (1994), Atuahene-Gima (1995) but inconsistent with the study of Greenley's (1995). It appears that the firms that commit their resources to discover additional technical features, desired by the customer, are more driven by new product development initiatives than their less market-oriented counterparts. Thus the relationship between market orientation and product-company fit is based on the assumption that a firm formulates and implements a specific competitive strategy to maximize the values of its capabilities (Porter, 1980). Therefore, new products that fit the firm's technological skills and resources are more likely to be successful (Cooper and Kleinschmidt, 1987). Further, there is emerging evidence that many engineering units, which are sources of radical innovations, are increasingly becoming market-oriented in the new product development process Workman (1993).

Market orientation is significantly ($p < 0.05$) and positively ($\beta = 0.20$) related to customer retention. Therefore, the hypothesis H1e that the market orientation is positively related to customer retention is supported. This finding is consistent with the study of Narver and Slater (1990). It appears that market-oriented companies have substantial control over their customers. In the machine tool sector, it was learned during the interview that the market-oriented companies were more successful in retaining their customers by providing them with customer-specific solutions.

Although attention to the needs of individual customers raised the cost of serving customers, market-oriented companies could record the impact of the competitive advantage on the their performance of companies in the long-term. It is not surprising that many market-oriented managers worked closely with their key account customers to form mutually profitable partnerships to retain customers (Wilson, 1992). Our finding appears to suggest that forming partnerships does not only bring repeat business and retain customers, but also it nurtures a relationship between the firm and the customer which over a period a time leads to their sense of belonging or commitment to organizations. Indeed, it is relatively easier for SMEs to cultivate such relationships with customers because of their specialised expertise that led to the formation of such a partnership.

Market orientation is significantly ($p < 0.00$) and positively ($\beta = 0.31$) related to the global presence; hence, the hypothesis H1f that the market orientation is positively related to global presence is supported. It appears that market-oriented companies have a policy of buying certain components from low-cost foreign suppliers, which does not only reduce the overall costs of machine but also help them access foreign markets by gaining new experience in the market. The fact that the companies practice the strategy of lowering production costs through foreign vendors indicates that these manufacturers already understand their markets, which give them an opportunity to develop a strategy to enter foreign markets by establishing subsidiaries or joint venture in less developed countries. For example, establishing subsidiaries can be cost effective by taking advantage of inexpensive labor and a segmented market. Western firms which are skilled at market-oriented activities effectively are likely to be viewed as attractive joint venture partners for foreign firms that recognize the importance of effective market-oriented practices but are under-performing in certain areas. Further, most technical specifications of machine tools are almost the same worldwide, so if a company is market oriented domestically, it is likely that it would have its presence abroad because of the machine's ability to perform similar operations. Therefore, it is reasonable to believe that market orientated firms should have their presence in foreign markets.

Effects of Control Variables on Market Orientation

With regard to the effects of control variables on performance, a mixed support for our expectations was observed. Competitor concentration and customer power were found to be significantly and negatively ($\beta = -0.23$, $p < 0.00$, $\beta = -0.18$, $p < 0.05$, respectively) related to market share. This suggests that as competitors' concentration and customer power increases, market share decreases. This finding is consistent with the received wisdom. Relative size is significantly ($p < 0.00$) and positively ($\beta = 0.39$) related to return on investment, significantly ($p < 0.05$) and positively ($\beta = 0.25$) related to sales growth, significantly ($p < 0.05$) and positively ($\beta = 0.24$) related to market share, significantly ($p < 0.01$) and positively ($\beta = 0.29$) related to new product success and significantly ($p < 0.01$) and positively ($\beta = 0.22$) related to the overall performance. These results indicate that business performance is improved when firms are larger than their competitors. The superior performance of companies may be attributed to the fact that large firms have more resources and experience, and therefore have more ability to perform better and achieve more economies of scale than their counterparts.

Relative cost is significantly ($p < 0.00$) and negatively ($\beta = -0.36$) related to return on investment, significantly ($p < 0.05$) and negatively ($\beta = -0.22$) related to new product success, significantly ($p < 0.01$) and negatively ($\beta = -0.29$) related to overall performance, and significantly ($p < 0.01$) and negatively ($\beta = -0.35$) related to global presence, implying that return on investment, new product success, overall performance and global presence are enhanced as costs relative to competitors reduce. Although, the ease of entry and the dominance of sales were not found to be statistically significant, their hypothesized signs are same as expected. Table 4.2 suggests that high costs of entry and high dominance of well-established companies might have rendered these relationships to be nonsignificant.

Analysis of Residuals

Following discussions about the assumptions relating to dependent variables, it is important to examine the residuals as they represent the error terms associated with the regression analysis. For the study, the following three kinds of analyses were performed on the residuals to validate the findings from the regression model.

Outliers: If the variance in an observation is not accurately explained by the regression equation, it generates an error associated with the corresponding observation, which may reduce the overall accuracy of the model. To identify outliers, a box plot (not shown) was produced for each dependent variable, which revealed that except for return on investment, new product success, overall performance and global presence, there were no residual outliers. The return on investment, new product success, overall performance and global presence have a maximum of three outliers. Given the ordinal nature of data and the small number of outliers, the regression model can be assumed to be acceptable.

Normal distribution of residuals: Another assumption relating to the regression model is that residuals should follow a normal distribution curve. If residuals do not conform to the normal distribution function, they may contain outliers that may distort the values of the regression parameters. To check if residuals were normally distributed, a normal P-P graph for standard residuals for each dependent variable was plotted. The diagonal straight line on the graph represents the normal distribution of the data. If residuals follow the line closely, they are assumed to follow a normal distribution curve. A K-S test for the normality of the residuals confirmed an acceptable departure of standard residuals from normality.

Equality of variance: This refers to the requirement that the error terms must have equal variance for different values of the dependent and independent variables. To test the equality of variance, residuals were standard and were plotted against corresponding standard predicted values. If the spread of residuals increases or decreases with the values of predicted variables, it indicates that the assumption of equality-of-variance is violated. From the data, it appeared that the residuals are within the acceptable range of equality of variance.

The Moderators

Although multiple regression analysis is a suitable technique to determine the degree of association between a set of predictors and criterion variables, results from several studies suggest that under certain circumstances the classic regression model may not provide a complete understanding of the phenomenon being studied. In some cases, it is possible that the predictive efficacy of an independent variable may vary systematically as a function of some other variable. For example, a classic regression model may not detect a significant relationship between A and B, but the relationship between the variables (A and B) may be significant in the presence of some other variable C. In other words, the strength of variable C may alter the relationship between variables A and B if all variables are considered together in the regression model.

As an alternative to the classical regression model, Saunders (1956) proposed the concept of moderator variable in the psychological literature. A moderator is a variable, which systematically modifies either the form or strength of the relationship between predictor and criterion variables.

While most researchers agree with the concept of a moderator variable, there is a substantial debate as to what exactly a moderator variable is and how it operates in influencing the classic validation model. For example, some researchers have stated that a variable is a moderator if it interacts with predictor variable (Fry, 1971; Horton, 1979; Peters and Champoux, 1979) irrespective of whether the hypothesized moderator is a significant predictor. A second concept is that a moderator cannot be a significant predictor variable nor can it be related to other predictor variables (Cohen and Cohen, 1975; Zedeck, 1971). Another notion is to ignore the interaction controversy by using an analytical procedure that examines the differences between individuals grouped on the basis of some hypothesized moderator variable (Bennet and Harrell, 1975; Ghiselli, 1960, 1963; Hobert and Dunnette, 1967). Sharma et al. (1981) propose two types of moderator variables. One type of moderator variable influences the model by affecting the strength of the relationship while the other modifies the form of the relationship.

In this study, I follow the typology suggested by Rosenberg (1968) and the framework offered by Sharma et al. (1981) for the detection and type of moderator variables. Such typology of specification variables and the framework for identifying moderator variables are presented in detail in the subsequent sections.

Typology of Moderators Variables

A moderator variable can be considered as a subset of a class of variables, which specifies (hypothesized variable is termed as specification variable until it is detected as one of the moderators or exogenous variables) the form or magnitude of the relationship between the predictor and the criterion variables (Rosenberg, 1968). Hence, specification variables can be mapped on two-dimensional characteristics. The first dimension can be based on the relationship with the criterion variable, i.e. whether or not the specification variable is related to criterion variable. The second dimension is whether or not the specification variable interacts with the predictor variable. Table 4.3 presents the typology of specification variables.

Table 4.3 Typology of specification variables

	Related to Criterion and/or Predictor	Not related to Criterion and/or Predictor
No interaction with Predictor	Intervening, Exogenous, Suppressor, Predictor	Moderator (Homologizer)
Interaction with Predictor variable	Moderator (Quasi Moderator)	Moderator (Pure Moderator)

Source: Rosenberg (1968)

The Framework for Identifying Moderator Variables

The proposed framework by Sharma et al. (1981) for identifying moderator variables consists of four steps. The graphical representation is shown in Figure 4.1.

Step one determines whether a significant interaction term is present between the hypothesized moderator variable, Z and the predictor variable as determined by the Moderated Regression Analysis (MRA) procedure. If a significant interaction is found, proceed to Step 2 otherwise go to Step 3.

Step two determines whether Z is related to the criterion variable. If so, Z is a quasi moderator variable, otherwise Z is a pure moderator variable. In either case, the moderator influences the form of the relationship in the classic validation model.

Step three determines whether Z is related to the criterion or predictor variable. If it is related, Z is not a moderator but an exogenous, predictor, intervening, antecedent, or a suppressor variable. If Z is not related to either the predictor or the criterion variable, proceed to Step four.

Step four splits the total sample into subgroups on the basis of a hypothesized moderator variable. Data can be split into groups by using certain criterion, such as mean, median, quartile, etc. After segmenting the total sample into the subgroups, do a test of significance for difference in predictive validity across subgroups. If significant differences are found, z is a homologizer variable operating through the error term. If no significant differences are found, z is a not a moderator variable and the analysis concludes.

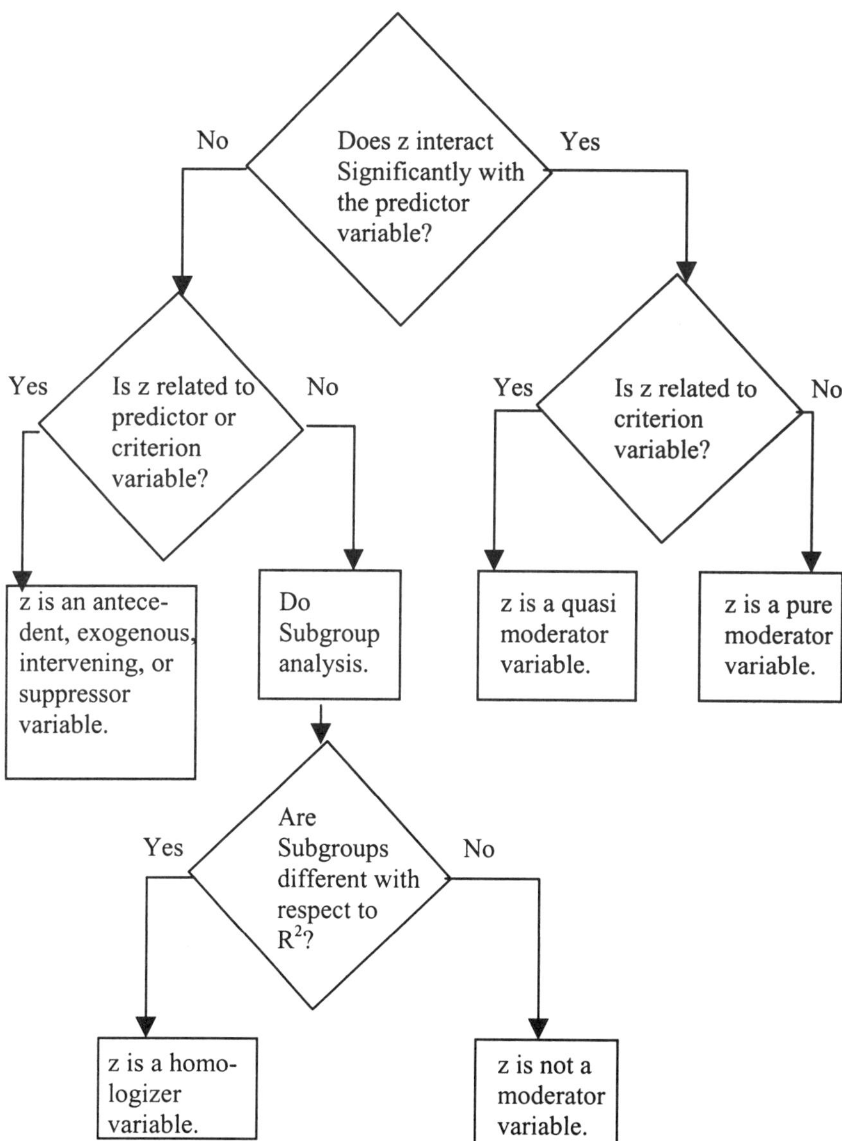

Figure 4.1 The framework for identifying moderators

Source: Sharma et al. (1981)

This study utilized three moderator variables - Market Turbulence, Competitive Intensity and Technological Turbulence - to examine if these moderatos altered the strength or form of relationship between market orientation and business performance. In the following section, construction of the moderators is explained along with corresponding statistical analyses.

Construction of Moderator Variables

Market Turbulence (MT): This refers to the extent to which customers' needs have changed over time and marketing operations have changed as a consequence to meet those changes (Miller, 1987). This variable consisted of four items (Jaworski and Kohli, 1993) and was measured on a seven-point scale, where one represented strongly disagree and seven strongly agree. Reliability analysis revealed that item four - we cater for many of the same customers that we used to in the past - was negatively correlated with the rest of the items. So it appears that firms are not successful in retaining customers may be attributed to the fact that companies are not able to provide what customers want, forcing customers to change suppliers. Lack of resources or innovation could be another possible reason for losing customers to competitors. To achieve high reliability, item four was eliminated. Reliability statistics have been reported in Table 4.4. The values of Cronbach Alpha and standard Alpha are 0.63 and 0.63, suggesting that the market turbulence scale may not be reliable.

Table 4.4 Market turbulence scale and reliability statistics

1. In our kind of business, customers' product preferences change quite a lot. (W/JK)
2. We are witnessing demand for our products and services from customers who never bought them before. (JK)
3. New customers tend to have products related needs that are different from those existing customers. (JK)
4. We cater for many of the same customers that we used to in the past. (JK)

Item	Corrected item-total correlation	α if item deleted
1	0.20	0.63
2	0.31	0.64
3	0.26	0.63

α = 0.63, Standard α = 0.63

Competitive Intensity (CI): It reflects the behaviour, resources and ability to differentiate itself from competitors (Jaworski and Kohli, 1993). The three-item variable was measured by borrowing items from Kohli and Jaworski (1993) on a seven-point scale, where one represented strongly disagree and seven strongly agree. These items with their reliability results are reported in Table 4.5. The scores for Alpha and standard Cronbach Alpha are above recommended level.

Table 4.5 Competitive intensity scale and reliability statistics

1 Competition in our industry is cut throat. (JK)
2 In our industry, anything that one competitor can offer, others can match readily. (W/JK)
3 Price competition is a hallmark of our industry. (JK)

Item	Corrected item-total correlation	α if item deleted
1	0.64	0.83
2	0.71	0.77
3	0.77	0.71

α = 0.83, Standard α = 0.84

Technological Turbulence (TT): This refers to the extent to which there is a change in magnitude in the technology associated with products or services and with research and development (Bennett and Cooper, 1981). The two-item scale (Jaworski and Kohli, 1993) was measured on a seven-point Likert scale, one being strongly disagree and seven being strongly agree. Table 4.6 indicates that values for Cronbach Alpha and standard Cronbach Alpha are 0.75 and 0.75, respectively, suggesting that the scale is reliable.

Table 4.6 Technological turbulence scale and reliability statistics

1 The technology in our industry is changing rapidly. (JK)
2 Technological changes provide opportunities in our industry. (JK)

α = 0.75, Standard α = 0.75

Subgroup Analysis

To test the moderating effects from the market turbulence, competitive intensity and technological turbulence on the market orientation-business performance relationship, three separate regression equations containing interaction term were estimated. It was found that none of the interaction terms was significant at $p < 0.05$. Of these criterion variables (ROI, SG and NPS) only two (ROI and NPS) were nonsignificantly related to the predictor variable (Market turbulence (MT) and technological turbulence (TT)). Further, MT and TT variables were not significantly related to the market orientation variable. This condition states that MT and TT variables may be treated as pure moderators, whereas the other hypothesized moderator variable (competitive intensity), which is significantly related to the criterion (Sales Growth), it may be considered as one of the following: antecedent, exogenous, intervening or suppressor variable. Because I did not have theoretical reason to consider the competitive intensity variable as antecedents, exogenous, intervening, or a suppressor, I decided to proceed with the subgroup to develop further insight into the relationship between competitive intensity, sales growth and market orientation. Table 4.2 suggests that nonsignificant correlation between return on investment and market turbulence, and between new product success and technological turbulence is 0.03 and -0.12, respectively, whereas correlation between sales growth and competitive intensity is statistically significant, $\beta = -0.37$ at $p < 0.00$. Following the framework of Sharma et al. (1981), a subgroup analysis was performed for the identification of moderators. Details of the subgroup analysis are presented in the subsequent sections.

The Approach to the Subgroup Analysis

The first step was the estimation of three regression equations involving multiple interaction term on the full sample to examine if the interaction term (β_3) was statistically significant. The constructed equation was as follow.

$$Y = \text{Constant} + \beta_1{*}MO + \beta_2{*}HMV + \beta_3{*}MO{*}HMV + \ldots + \beta_n{*}X_n$$

Where, Y=Dependent Variable, MO=Market Orientation, HMV=Hypothesized Moderator Variable, β_n=Partial Regression Coefficients.

The second step was to divide the sample by the medians of the hypothesized moderator variables, market turbulence, competitive intensity, technological turbulence, into two LO and HI subgroups with the purpose of determining weather the associations between market orientation and business performance differed significantly between the two subgroups.

The third step was to build a separate regression model on the subgroups considering ROI, SG and NPS as dependent variable and the rest of the variables as independent variables. Care was taken to reduce multicollinearity by checking the outliers in the independent variables. Preliminary results suggest that there are significant differences between the βs, which was calculated by using Fisher's Z-test.

The next step was to estimate regression equations on the subgroups pooled together. Of primary interest was the sum of squared errors for the pooled sample (SSEp) and the extent to which that value differed from the errors obtained from the subgroup regressions. If the errors for the subgroups were small relative to the errors for the pooled sample, the best fitting models for the two subgroups differed from one another. The specific form of the Chow (1960) test was applied for the detection of the strength of the hypothesized variable. F-static was calculated by using the formula suggested by Hambrick and Lei (1985).

$$F_{(k,\, n_1+n_2-2k-2)} = \frac{[SSE_p - (SSE_1 + SSE_2)] / k}{[(SSE_1 + SSE_2) / (n_1 + n_2 - 2k - 2)]} \qquad [\text{Eq}^n\ 4.1]$$

Where, SSE_p=Sum of squared errors for pooled samples, SSE_1=Sum of squared errors for subgroup 1, SSE_2=Sum of squared errors for subgroup 2, n_1=Size of subgroup 1,

n_2=Size of subgroup 2 and k=Number of independent variables in the model.

Effect of market turbulence as hypothesized moderator between market orientation and return on investment: In the pooled sample, the following regression equations were estimated. Equations 4.2 and 4.3 are without and with multiple interaction term, respectively.

$$ROI = Constant + \beta_1*MO + \beta_2*MT + ...+ \beta_n*X_n \quad [Eq^n\ 4.2]$$
$$ROI = Constant + \beta_1*MO + \beta_2*MT + \beta_3*MO*MT + ...+ \beta_n*X_n \quad [Eq^n\ 4.3]$$

Where ROI=Return on Investment, MO=Market Orientation, MT=Market Turbulence (the hypothesized moderator variable), MO*MT=Multiplicative Interaction Term (MIT), X_n=Independent Variables, β_n=Partial Regression Coeff.

Table 4.7 Partial regression coefficients (standard error)

		MIT/ROI	LO/ROI	HI/ROI
H2a	MT	-0.15 (0.14)	0.57^{**} (0.13)	$-0.21^{*\#}$ (0.06)
		MIT/NPS	LO/NPS	HI/NPS
H2b	TT	-0.04 (0.08)	0.53^{**} (0.10)	$0.23^{*\#}$ (0.05)
		MIT/SG	LO/SG	HI/SG
H2c	CI	0.08 (0.05)	$-0.41^{*\#}$ (0.13)	-0.19^{*} (0.16)

$^{*, **}$coefficients significant at $p < 0.05$ and $p < 0.01$, respectively
$^{\#}$ partial correlation coefficients different at $p < 0.01$, employing Fisher's Z-test (Hambrick and Lei, 1985). MIT=Multiplicative Interaction Term, LO=low, HI=high, ROI=Return on Investment, NPS=New Product Success and SG=Sales Growth.

Table 4.8 HMV: Market turbulence, and DV: ROI

	Pooled sample (p), without MIT	Full sample, with MIT	HI sample (1), without MIT	LO Sample (2), without MIT
SSE	62.47	71.98	19.99	27.40
F	8.14	6.84	6.68	7.86
R^2	0.52	0.56	0.60	0.65
Adj. R^2	0.45	0.53	0.55	0.57
VIF_{max}	1.86	1.98	2.95	2.32
k	10	10	8	8
n	91	91	44	42

Where, DV=Dependent Variable, HMV=Hypothesized Moderator Variable, MIT =Multiplicative Interaction Term, k= No. of Independent Variables, n= Sample Size.

Because the market turbulence variable was hypothesized as a moderator, this variable was split at the median into two groups: low market turbulence and high market turbulence groups. Median for the variable is 4.5 on a seven-point Likert scale. Next, the regression parameters for Equation 4.2 were estimated on both low and high market turbulence groups. Table 4.7 reports the β_s for both low and high market turbulent subgroups. Sum of squared errors for pooled samples (without interaction term), low and high market turbulent subgroups, and for pooled sample with interaction term have been reported in the Table 4.8.

Substituting the values of the parameters in the Equation 4.1, as proposed by Chow (1960), the following value for the F-static was obtained.

$$F_{(8,\ 44+42-2*8-2)} = \frac{[62.47-(19.99+\ 27.40)]\ /\ 8}{[(19.99+27.40)\ /\ (44+42-2*8-2)]}$$

$$F_{(8,68)} = \frac{1.885}{0.658},\quad F_{(8,68)} = 2.86$$

The value of F for (8,68) is greater than 2.78 (for 8,68 and at $p < .01$, using the F-statistic table from Cohen and Cohen, 1975), suggesting that there are significant differences in the association between market orientation and return on investment in businesses operating under low and high market turbulent environments. Table 4.7 shows the nonsignificant coefficients for the multiplicative interaction terms; however, the differences in the magnitude of the partial correlation coefficients of the market orientation between the low (0.57, $p < .01$) and high (-0.21, $p < 0.05$) market turbulence are significant for the return on investment. The result indicates that the market orientation has a more positive impact on return on investment in case of low market turbulence. Further Table 4.2 suggests that the market turbulence is not significantly (0.03, $p < 0.05$) related to the return on investment. Therefore, market turbulence can be considered as a homologizer moderator, which alters the form of the relationship between market orientation and return on investment. Therefore, the hypothesis H2a that the lesser the extent of market turbulence, the greater the positive impact of market orientation on performance (ROI) is supported.

Effect of technological turbulence as a hypothesized moderator between market orientation and new product success rate: Regression Equations 4.4 and 4.5 were built without interaction term and with interaction term, respectively in the full sample.

$$NPS = Constant + \beta_1*MO + \beta_2*TT + ... + \beta_n*X_n \qquad [Eq^n\ 4.4]$$
$$NPS = Constant + \beta_1*MO + \beta_2*TT + \beta_3*MO*TT + ... + \beta_n*X_n \qquad [Eq^n\ 4.5]$$

Where NPS=New Product Success Rate, MO=Market Orientation, TT=Technological Turbulence (the hypothesized moderator variable), MO*TT=Multiplicative Interaction Term (MIT), X_n=Independent Variables, β_n=Partial Regression Coefficients.

Technological turbulence was split by its median into two groups: low technological turbulence and high technological turbulence groups. Median for the variable is 5.5 on a seven-point Likert scale. Regression parameters for Equation 4.4 were estimated in both low and high technological turbulence subgroups. Table 4.7 reports the values for the β_s in the low and high technological turbulent subgroups. Sum of squared errors for pooled samples, low and high technological turbulent subgroups, and for pooled sample with interaction term have been reported in Table 4.9.

Table 4.9 HMV: Technological turbulence, and DV: new product success

	Pooled sample (p), without MIT	Full sample, with MIT	HI sample (1), without MIT	LO Sample (2), without MIT
SSE	39.98	35.63	14.70	13.34
F	8.58	9.80	3.06	8.69
R^2	0.56	0.58	0.48	0.62
Adj. R^2	0.47	0.53	0.32	0.55
VIF_{max}	1.88	1.70	3.30	2.03
k	10	10	8	8
n	85	85	35	50

Where DV=Dependent Variable, HMV=Hypothesized moderator variable, MIT=Multiplicative Interaction Term. As Chow (1960) proposed, F-static was calculated by substituting the values of the parameters in the Equation 4.1.

$$F_{(8,\ 35+50-2*8-2)} = \frac{[39.98 - (14.70 + 13.34)] / 8}{[(14.70 + 13.34) / (35 + 50 - 2*8 - 2)]}$$

$$F_{(8,67)} = \frac{1.49}{0.39},\ F_{(8,67)} = 3.82$$

The value of F for (8,67) is greater than 2.78 (for 8,67 and at $p < 0\ .01$), indicating that there are significant differences in the correlations between market orientation and new product success in businesses functioning under the low and high conditions of technological turbulences. Table 4.7 suggests that the coefficient for the multiplicative interaction term is statistically insignificant; however, the differences in the magnitude of the partial correlation coefficients of the market orientation between the low (0.53, $p < 0.01$) and the high (0.23, $p < 0.05$) technological turbulences are significant for new product success. These results suggest that the market orientation appears to have a more positive impact on new product success when technological turbulence is low. As Table 4.2 indicates that the technological turbulence is not significantly (-0.12, $p > 0.05$) related to the new product success, it may be concluded that technological turbulence can be considered as a homologizer moderator, which alters the form of the relationship between the market orientation and the new product success. Therefore, the hypothesis H2b that the lesser the extent of technological turbulence, the greater the positive impact of market orientation on performance (NPS) is supported.

Effect of competitive intensity as hypothesized moderator between market orientation and sales growth: Similarly, in the full sample, the following two regression equations were estimated.

$$SG = \text{Constant} + \beta_1 {*} MO + \beta_2 {*} CI + \ldots + \beta_n {*} X_n \quad [Eq^n\ 4.6]$$
$$SG = \text{Constant} + \beta_1 {*} MO + \beta_2 {*} CI + \beta_3 {*} MO {*} CI + \ldots + \beta_n {*} X_n \quad [Eq^n\ 4.7]$$

Where SG=Sales Growth, MO=Market Orientation, CI=Competitive Intensity (hypothesized moderator variable), MO*CI=Multiplicative Interaction Term (MIT) X_n=Independent Variables, β_n=Partial Regression Coefficients.

Similarly, competitive intensity was split by its median into two groups: low competitive intensity and high competitive intensity groups. Median for the variable is 4.67 on a seven-point scale. Regression parameters for Equation 4.6 were estimated for both the low and high competitive intensity subgroups. Table 4.7 reports the β coefficients for the low and high competitive intensity subgroups. Sum of squared errors for pooled samples, LO and HI competitive intensity samples, and pooled sample with interaction term have been reported in the Table 4.10.

Table 4.10 HMV: Competitive intensity, and DV: sales growth

	Pooled sample (p), without MIT	Full sample, with MIT	HI sample (1), without MIT	LO Sample (2), without MIT
SSE	20.98	17.03	10.30	3.96
F	6.24	7.04	7.98	5.03
R^2	0.48	0.51	0.71	0.53
Adj. R^2	0.40	0.47	0.62	0.49
VIF_{max}	1.71	2.68	3.11	2.90
k	10	10	8	8
n	78	78	44	44

Where, DV=Dependent Variable, HMV=Hypothesized Moderator Variable, MIT =Multiplicative Interaction Term. Similarly, substituting the values of the parameters in the Equation 4.1, the following value of F-static was calculated.

$$F_{(8,\ 44+44-2*8-2)} = \frac{[20.98 - (10.30 + 3.96)] / 8}{[(10.30 + 3.96) / (44 + 44 - 2*8 - 2)]}$$

$$F_{(8,70)} = \frac{0.84}{0.19}, \; F_{(8,70)} = 4.42$$

The value of F for (8,70) is greater than 2.74 (for 8,70; $p < 0.01$), suggesting that there is a significant difference in the association between the market orientation and the sales growth in companies operating under environments that are characterised by low or high competitive intensity. Table 4.7 shows that the coefficient for the multiplicative interaction term is statistically nonsignificant ($p > .05$); however, the differences in the magnitude of the partial correlation coefficients of market orientation between the low (-0.41, $p < 0.05$) and the high (-0.19 $p < 0.05$) competitive intensity environments are significant.

The results indicate that the market orientation has less negative impact on sales growth when competitive intensity is high. Further, Table 4.2 suggests that competitive intensity is significantly (-0.37, $p < 0.00$) related to the sales growth. Competitive intensity may be assumed to be intervening variable. Hence, the hypothesis H2c that the greater the degree of competitive intensity, the greater the positive impact of market orientation on performance (SG) is partially supported.

Conclusions: Moderator Effects Based on the Subgroup Analysis

Our results indicated that relationship between market orientation and business performance in three instances: return on investment is high when market turbulence is low; new product success is high when technological turbulence is low; and sales growth is high when competitive intensity is high.

Market turbulence (H2a): Our finding that the lesser the extent of market turbulence, the greater the positive impact of market orientation on performance is consistent with the studies of Slater and Narver (1994) and Greenley (1995). In line with expectations, these findings reveal that market orientation may be more relevant in less turbulent market conditions. Therefore, it seems that in such market conditions, successful firms are those who pay more attention to the changing market needs to satisfy customer preferences. In fact, a substantially market-oriented business should find more opportunities in any environment than its less market-oriented counterparts (Kirzner, 1979). This is why market orientation is as important, if not more important, during low market turbulence as it is during high market turbulence (Slater and Narver, 1994). In the context of the machine tool industry where the market is highly cyclical in nature and dependent upon the strength of the economy, it is logical to think in terms of the cost and the benefits of increasing market orientation. This is because the costs associated with the changing marketing operations become disproportionately high relative to sales revenue at high levels of market turbulence, and market conditions may be such that sales revenue cannot be increased, resulting in a decline in profits (Greenley, 1995). Therefore, in highly turbulent markets it may be preferable to impose a consistent approach to marketing operations, rather than responding erratically (e.g., new product development, major changes on existing product) to the changing market. Becoming more market-oriented may result in more costs and fewer benefits.

Technological turbulence (H2b): Our finding that the lesser the extent of technological turbulence, the greater the positive impact of market orientation on performance is consistent with the studies of Slater and Narver (1994) and Greenley (1995). It appears that under low technological turbulence, the association between market orientation and new product success is stronger. There seems to be a turning point where, beyond a certain level of technological innovation, additional customer benefits from technology cannot be achieved. Therefore, improving market orientation by introducing a new product during a high level of technological change may be ineffective (Greenley, 1995). In technological turbulent conditions, customer preferences cannot detect the evolution of new products in a timely fashion. It is particularly true in the machine tool industry because the advancement in technology is so rapid that by the time a new product is ready to be launched, the technology may become obsolete, rendering the product commercially less successful. Most small businesses are probably not affected to a great extent by such changes, because they maintain a consistent focus. Indeed, many small firms may not have the resources to introduce radical changes to gain the sustainable competitive advantage (Narver and Slater, 1990).

Competitive intensity (H2c): Our finding that the greater the degree of competitive intensity, the greater the positive impact of market orientation on performance is consistent with the received wisdom (Diamantopolous and Hart, 1993; Atuahene-Gima, 1995; Appiah-Adu, 1998). This implies that at low levels of competitive intensity, machine tool firms do not devote much attention and resources to enhancing their degree of market orientation. This may be due to the fact that the product has: (a) a unique attribute which sells machine; (b) there is a monopoly in the market or that the product features are too expensive or too complicated to copy which reduces the need for increasing market orientation. On the other hand, if there are several players in the market with almost similar characteristics, there is a pressing need for becoming market oriented by paying closer attention to the customer needs and providing a unique or total solution to the problem. Our results have shown that higher competition leads to a higher sales growth. Certainly, if firms are able to provide custom-made solutions to their customers, it is likely to increase the sales of the company. Further, the culture of being market-oriented may give rise to customer retention. A satisfied customer may come to the same firm seeking solution for another problem. Hence, a problem solving strategy by providing total solution to customers may help retain customers, which in turn, might lead to higher sales growth.

Moderated Regression Analysis

Table 4.2 contains the results of the main effects of market orientation and independent variables on return on investment, sales growth, market share, new product success, customer retention, overall performance and global presence. Market orientation is the only predictor whose regression coefficient is nonsignificant (0.16, $p < 0.05$), suggesting that there is no main effect from market orientation on overall performance. Although an association was not identified in the main analysis, the effects from moderator variables may be present, which can be detected by employing moderated regression analysis (Schooven 1981; Sharma, Durand and Gur-Arie, 1981; Arnold 1982; Hellivik 1984; Golden, 1992).

The hypotheses H2a, H2b and H2c were tested in which overall performance of companies was considered as a business performance indicator.

Approach to Moderated Regression Analysis

A pure moderating effect implies that the moderator modifies the form of relationship between predictor and criterion variables. The Moderated regression analysis involves estimation of the equation with a multiplicative interaction term in the total sample. It does not require explicit subgroup of variables. The R^2 determines whether the multiplicative term in the equation contributes to prediction beyond that of the ordinary linear regression, i.e. the equation without the multiplicative interaction term. A detailed moderated regression analysis is presented in the following sections.

Two regression equations: one without the multiplicative interaction term and other with the multiplicative interaction term, were estimated to compare the variance explained by each equation.

$$Y = \text{Constant} + \beta_1{*}MO + \beta_2{*}HMV + ... + \beta_n{*}X_n \qquad [Eq^n 4.8]$$
$$Y = \text{Constant} + \beta_1{*}MO + \beta_2{*}HMV + \beta_3{*}MO{*}HMV + ... + \beta_n{*}X_n \qquad [Eq^n 4.9]$$

Where, Y=Dependent Variable, MO=Market Orientation, HMV=Hypothesized Moderator Variable, MO*HMV=Multiplicative Interaction Term (MIT), β_n=Partial Regression Coefficients.

A moderating effect is detected when the regression coefficient (β_3) of the multiplicative interaction term is statistically significant, in addition to the significance of the coefficients of market orientation (β_1) and the hypothesized moderator variable (β_2). A further analysis can identify any difference in the form of the relationship between the predictor variable (market orientation) and the dependent variable (overall performance), over a range of moderator variables (market turbulence, technological turbulence and competitive intensity). For more information about deriving equations that represent the relationship between independent variables and dependent variables at various levels of moderators, see Aiken and West (1991) and Turrissi and Jaccard (2003). For the purpose of the study, moderators can be determined by taking the partial derivative of the Equation 4.9 with respect to the market orientation.

$$\frac{\partial Y}{\partial MO} = \beta_1 + \beta_3 * HMV$$

The Mean Centred Correction for MRA models

Although the evaluation of multiplicative term is appropriate for interval data, the use of ordinal data is acceptable as well in marketing research. Busemeyer and Jones (1983) highlight the biased effects of departure even from interval data. While it is true that the regression analysis should be usually conducted with interval data, ordinal data may be regressed if such data approximate interval data characteristics (Birnbaum, 1982). In this study, multiple regression analysis was employed to analyze the ordinal data as this technique has been used widely by researchers for the detection of main and moderating effects between criterion and predictor variables (Kohli and Jaworski, 1990; Slater and Narver, 1994; Greenley, 1995, Pitt et al., 1996).

A caution against the use of multiplicative interaction term in regression analysis focuses on the multicollinearity. Althauser (1971) and Blalock (1979) have noted that multiplicative terms usually exhibit strong correlations with predictors; therefore, introducing inflated standard errors. Because multiplicative terms can introduce high level of multicollinearity in moderated regression equation, Jaccard et al. (1990) suggest that all predictor variables should be mean centred before multiplicative interaction term is formed to reduce the effects of multicollinearity and standard error.

Consistent with Jaccard's (1990) suggestion, all predictor variables were mean centred by subtracting the mean value (4) from each case on the predictor variables. The mean-centred scores were then used in the final analysis. A multiple regression analysis was then conducted regressing dependent variable on independent variables and the multiplicative interaction term using the SPSS software. A detailed analysis for each individual moderator variable is explained in the subsequent sections.

The Individual Moderated Regression Analysis

Effect of the market turbulence as a hypothesized moderator between the market orientation and overall business performance (p-ov): In the full sample, the following regression equations were estimated: Equation 4.10 is the ordinary linear regression equation whereas Equation 4.11 contains the multiplicative interaction term (MO*MT). Table 4.11 reports the β_s coefficients and other related statistics to the moderated regression equations.

$$\text{p-ov} = \text{Constant} + \beta_1 * MO + \beta_2 * MT + ... + \beta_n * X_n \qquad [Eq^n 4.10]$$
$$\text{p-ov} = \text{Constant} + \beta_1 * MO + \beta_2 * MT + \beta_3 * MO * MT ... + \beta_n * X_n \qquad [Eq^n 4.11]$$

Where, p-ov=Overall Performance, MO=Market Orientation, MT=Market Turbulence (hypothesized variable), MO*MT = Multiplicative Interaction Term (MIT), β_n=Partial Regression Coefficients, X_n=Independent Variables.

Taking the overall performance (P-ov) as a dependent variable and the market turbulence as a hypothesized moderator variable and substituting their partial regression coefficients from Table 4.11 in Equation 4.11, the following equation was estimated.

$$\text{p-ov} = 3.27 + 0.67 * MO + 0.23 * MT - 0.21 * MO * MT \qquad [Eq^n 4.12]$$

Table 4.11 Moderated regression analysis: market turbulence

	DV = p-ov (Overall performance)	
Independent variables	t	β
Market Orientation	2.46	0.67^{**}
Market Turbulence	2.37	0.23^{**}
Interaction term	-2.22	-0.21^{*}
R^2	0.75	
Adj. R^2	0.70	
F	16.10	
VIF_{max}	1.86	
Constant term	3.27	

$^{*}= p < 0.05$, $^{**} = p < 0.01$, $^{***} = p < 0.00$

The partial coefficients for the market orientation ($\beta = 0.67, p < 0.01$), the market turbulence ($\beta = 0.23$, $p < 0.05$) and the multiplicative term ($\beta = -0.21$, $p < 0.05$) were found to be significant. Tables 4.2 and 4.11 indicate that the difference between the variance in the main model and the model incorporating the multiplicative interaction term is (adjusted R^2 score) +9 per cent for the overall performance indicator, suggesting that the link between the market orientation and overall performance is stronger when an interaction term is included. The form of this moderation was examined by taking partial derivative of the Equation 4.12 with respect to the market orientation. Further equating it to zero, a point of influx was determined.

$$\frac{\partial p-ov}{\partial MO} = 0.67 - 0.21 * MT = 0 \qquad [Eq^n 4.13]$$

$$MT = \frac{0.67}{0.21} = 3.19$$

When scores for market turbulence in excess of 3.19 are substituted in the Equation 4.13, it yields negative values, while scores below 3.19 yield positive values. This indicates that for medium to high levels of market turbulence, market orientation is negatively associated with overall performance; however, market orientation is positively related to overall performance under conditions of low market turbulence. Therefore, hypothesis H2a that the lesser the extent of market turbulence, the greater the positive impact of market orientation on performance is supported. The finding from the moderated regression analysis is same as the finding from subgroup analysis, lending credence to the validity of the result.

Effect of technological turbulence as a hypothesized moderator between market orientation and business performance: Similarly, in the full sample, the following regression equations were estimated.

p-ov = Constant + β_1*MO + β_2*TT +...+ β_n*X_n [Eq^n 4.14]

p-ov = Constant + β_1*MO + β_2*TT +β_3*MO*TT +...+ β_n*X_n [Eq^n 4.15]

Where p-ov=Overall Performance, MO=Market Orientation, TT=Technological Turbulence, MO*TT=Multiplicative Interaction Term (MIT), β_n=Partial Regression Coefficient, X_n=Independent Variables.

Coefficients and the related statistics for the regression equations are reported in the Table 4.12.

Table 4.12 Moderated regression analysis: technological turbulence

	DV = p-ov (Overall performance)	
Independent variables	t	β
Market orientation	2.46	0.79*
Technological turbulence	-2.80	-0.31***
Interaction term	-2.39	-0.22*
R^2	0.69	
Adj. R^2	0.64	
F	15.26	
VIF_{max}	1.79	
Constant term	4.16	

* = $p < 0.05$, ** = $p < 0.01$, *** = $p < 0.00$

Considering the overall performance as a dependent variable and the technological turbulence as a hypothesized moderator and substituting their partial regression coefficients from Table 4.12 in Equation 4.15, the following equation was obtained.

p-ov = 4.16 + 0.79*MO - 0.31*TT - 0.22*MO*MT [Eq^n 4.16]

The partial coefficients for market orientation (β = 0.79, $p < 0.05$), the technological turbulence (β = -0.31, $p < 0.00$ and the multiplicative term (β = -0.22, $p < 0.01$) were found to be significant. Tables 4.2 and 4.11 indicate that the difference between the variance in the main model and the model incorporating the multiplicative interaction term is + 3 per cent for the overall performance variable. The improvement in the variance in the model including interaction term may seem rather marginal; nonetheless, these findings indicate that the form of relationship between market orientation and overall performance changes in the presence of the interaction term. Further, the form of the relationship was examined by taking partial derivative of the Equation 4.16 with respect to the market orientation. Equating the equation to zero, a point of inflection was determined.

$$TT = \frac{0.79}{0.22} = 3.59$$

$$\frac{\partial p-ov}{\partial MO} = 0.79 - 0.22 * TT = 0 \qquad [Eq^n 4.17]$$

Hence, when scores above 3.59 for the technological turbulence are substituted in Equation 4.17, it gives negative values, while scores below 3.59 yield positive values, indicating that for high levels of technological turbulence, the market orientation is negatively associated with the overall performance and for low levels of technological turbulence, market orientation is positively related to overall performance. Therefore, the hypothesis H2b that the lesser the extent of technological turbulence, the greater the positive impact of market orientation on performance is supported. The finding from this analysis is validated from the finding of subgroup analysis, as both methods delivered the same results.

Effect of competitive intensity as a hypothesized moderator between market orientation and business performance: Similarly, Table 4.13 reports the values for the β_s and other statistics for the following regression equations.

$$\text{P-ov} = \text{Constant} + \beta_1 * MO + \beta_2 * CI + ... + \beta_n * X_n \qquad [Eq^n 4.18]$$
$$\text{P-ov} = \text{Constant} + \beta_1 * MO + \beta_2 * CI + \beta_3 * MO * CI + ... + \beta_n * X_n \qquad [Eq^n 4.19]$$

Where p-ov=Overall Performance, MO=Market Orientation, CI=Competitive Intensity (hypothesized variable), MO*CI=Multiplicative Interaction Term (MIT), β_n=Partial Regression Coefficients, X_n=Independent Variables.

Table 4.13 Moderated regression analysis: competitive intensity

	DV = p-ov (Overall performance)	
Independent variables	t	β
Market Orientation	-2.35	-0.65^{*}
Competitive Intensity	-2.61	-0.23^{***}
Interaction term	3.79	0.19^{***}
R^2	0.72	
Adj. R^2	0.66	
F	14.47	
VIF_{max}	1.95	
Constant Term	2.76	

$^{*} = p < 0.05$, $^{**} = p < 0.01$, $^{***} = p < 0.00$

Using the overall performance variable as a dependent variable and the competitive intensity as a moderator variable and substituting their partial regression coefficients from Table 4.13 in Equation 4.19, the following regression equation was obtained.

$$\text{p-ov} = 2.76 - 0.65*\text{MO} - 0.23*\text{CI} + 0.19*\text{MO}*\text{CI} \qquad [\text{Eq}^n\ 4.20]$$

The partial coefficients for the market orientation ($\beta = -0.65$, $p < 0.05$), the competitive intensity ($\beta = -0.23$, $p < 0.00$ and the multiplicative term ($\beta = -0.19$, $p < 0.00$) were found to be significant. Table 4.2 and 4.13 indicate that the difference between the variance in the main model and the model incorporating the multiplicative interaction term is +5 per cent.

These findings indicate that the presence of the interaction term alters the relationship between market orientation and overall performance. By taking a partial derivative of Equation 4.20 with respect to the market orientation and equating it to zero, a point of inflection was determined.

$$\frac{\partial p - ov}{\partial MO} = -0.65 + 0.19 * CI = 0 \qquad [\text{Eq}^n\ 4.21]$$

$$CI = \frac{-0.65}{-0.19} = 3.42$$

From the equation an inflection point of 3.42 was obtained, which indicates that the equation yields a positive value when the scores for the competitive intensity above 3.42 are substituted in Equation 4.21. This means that the market orientation is positively associated with the overall performance in times of high competitive intensity, while for lower levels of competitive intensity, market orientation is negatively associated with the overall performance. Hence, the hypothesis H2c that the higher the extent of competitive intensity, the greater the positive impact of market orientation on performance is accepted. Same findings from subgroup and moderated regression analysis techniques lend credibility to the results.

Conclusions: Effects of Moderators on Market Orientation Business Performance Relationship

In the moderated regression analysis, the mean of the five performance indicators was used to examine the impact of the market orientation on the business performance under various environmental conditions: market turbulence, technological turbulence and competitive intensity. The findings from the subgroup analysis identified MT and TT as homoglozier moderators and CI as intervening variable, whereas moderated regression analysis identified MT as a quasi moderator and TT and CI as pure moderators. It may be noted that there is a slight difference in the coordinates of inflection points for the moderators in the two methods.

These differences were found because different performance indicators were used in different techniques. For example, in the subgroup analysis, market turbulence, technological turbulence and competitors intensity were hypothesized as moderators taking return on investment, new product success and sales growth as performance indicators, respectively, whereas in the case of moderated regression analysis, only overall performance was used as a performance indicator while keeping the moderators the same as in subgroup analysis.

For stronger association between the market orientation and the business performance, the value for the market turbulence was found to be less than 4.5 in case of subgroup analysis on return on investment, whereas it should be less than 3.19 in case of moderated regression analysis on the overall performance. Similar differences were observed for other moderators: technological turbulence and competitive intensity. The technological turbulence on new product success appeared to have a score less than 5.5 in subgroup analysis and less than 3.59 in the moderated regression analysis on the overall performance for the relationship to be significant between market orientation and business performance. Finally, the values for the competitive intensity was found to be more than 4.67 in the subgroup analysis on sales growth and more than 3.42 in the moderated regression analysis on the overall performance for a significant positive association between the market orientation and the business performance. Because the nature of the relationship remains the same in both techniques, it may be inferred that these findings are valid.

Parsimonious Models of the Market Orientation Factors

To test the impact of each factor of the market orientation on the business performance, a linear regression equation model was built using the Enter method. The statistics for the general model are presented in the Table 4.14.

Table 4.14 General model of the market orientation

Business performance = *f* (Factors underlying market orientation concept)

Factors	β	Std. Error	t	Significance
F 1 (Customer)	0.32	0.11	2.95	0.00
F 2 (Competitor)	-0.32	0.09	-2.84	0.00
F 3 (Responsive)	-0.05	0.09	-0.52	0.60
F 4 (Satisfaction)	0.49	0.10	3.93	0.00
F 5 (Market info.)	0.14	0.07	1.34	0.18
F 6 (Marketing)	0.21	0.07	2.14	0.03
F 7 (Negligence)	-0.09	0.06	-1.05	0.29
R^2	=0.47			
Adj. R^2	=0.42			
F	=10.83			

Development of the Parsimonious Market Orientation Model

To achieve a parsimonious model and to reduce the effect of multicollinearity, a regression analysis was performed in two stages. The first stage of the analysis considered seven factors as independent variables and the business performance as a dependent variable. The main purpose of the analysis was to detect the significant factors that accounted for the variance in the business performance variable. In the second stage, the business performance variable was regressed on factors which were found to be significant. This two-stage procedure enables development of a parsimonious model. The results of the parsimonious models are reported in Table 4.15, which suggests that the resulting parsimonious model accounts for 41.1 per cent of the variation in the business performance variable, which is approximately the same as the results from the model incorporating all seven factors. In other words, factors 3 (responsiveness), 5 (market information) and 7 (negligence) did not contribute significantly to enhancing the variance in the business performance. Statistics relating to the parsimonious model are presented in Table 4.15.

Table 4.15 Parsimonious model of business performance

Business performance = *f* (Factors underlying market orientation concept)

Factors	β	Std. Error	t	Significance
Customer	0.33	0.09	3.69	0.00
Competitor	-0.35	0.08	-3.20	0.00
Satisfaction	0.49	0.08	4.31	0.00
Marketing	0.24	0.06	2.71	0.00
R^2	=0.45			
Adj. R^2	=0.41			
F	=18.06			

The One Way Analysis of Variance of the Significant Factors

Although multiple regression analysis is a suitable technique for examining the relationship between a dependent variable and several independent variables, it is worthwhile to use the subgroup analysis to test the equality of means across subgroups. Each four factors were split into three mutually exclusive low (LO), medium (MI) and high (HI) subgroups. Cut-off values on these factors were selected in such a way that each group had almost the same number of respondents. Table 4.16 reports the results of the one-way ANOVA. For easy visual representation, the means are plotted in Figure 4.2.

Results and Discussions

Customer focus: Table 4.15 suggests that the customer orientation factor has a significant ($p < 0.00$) and a positive (0.33) effect on the business performance. More specifically, medium and high customer focus activities lead to more profitable business than a low customer focus (Table 4.16 and Figure 4.2). The machine tool industry generally has a set of three customers: basic machine tool users (non-CNC machine tools); sophisticated machine tool users (hand-held machine tools); and, very sophisticated machine tool users (CNC machine tools). Therefore, machine tool manufacturers usually have different customer orientation strategies for different sets of machine tool users. For example, companies manufacturing highly technologically advanced machines will devote more attention to customer focus than companies which manufacture basic machine tools. This implies that companies that are successful in creating a niche market perform relatively better as a result of being more customer-oriented. Therefore, it is vital for a company to cultivate a culture required to achieve and maintain superior performance of machines by developing high quality machines that are specific to customers' needs. Certainly, knowledge of customers' needs is paramount to the survival and growth of companies, particularly SMEs (Berkowitz et al., 2003). Although these companies claim to be customer-oriented, there are a few companies that truly understand the concept as the meanings of the concept varies from 'offering the customer what the customer wants' to 'understanding and solving customer problems'. It also makes sense for SMEs to develop good segmentation strategies by becoming specialist niche players to solve customers' problems and form relationship. As a result, this relationship can be extended to joint product testing, production of custom-made machines as well as diffusion of new products.

Table 4.16 Means of business performance

Focus	Customer	Competitor	Satisfaction	Marketing
LO	4.90*	5.42	5.14*	4.96*
MI	5.64*	5.34	5.38*	5.59*
HI	6.00*	5.73	6.02*	6.03*

* $p < 0.05$

Key: LO=Low, MI=Medium, HI=High

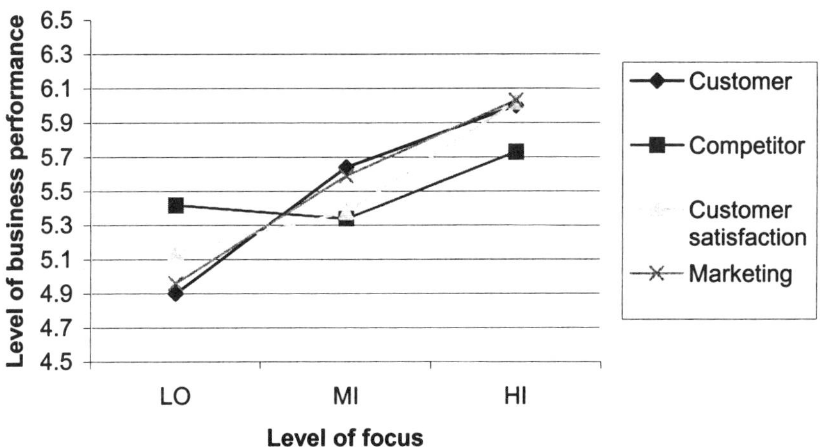

Figure 4.2 Level of focus vs. level of business performance

Key: LO=Low, MI=Medium, HI=High

Competitor focus: The focus has a significant ($p < 0.00$) and a negative ($\beta = -0.35$) effect on business performance. Table 4.16 indicates that low and high competitor focus activities contribute more to business performance than medium competitor focus activities. Although these differences are not significant but marginal in that the costs of becoming competitor oriented outweigh the benefits when the level of competitors' orientation is medium. These findings offer support for the view that companies, particularly original equipment manufacturer, become progressively less competitor oriented following the receipt of an order. This is because there is now more need to be customer oriented (to execute the order) than to be competitor oriented, as there are no further competitive activities till next competition for submission of tenders. The reduction in competitive activities is shown by the graph in Figure 4.2 with negative gradient while the customer orientation gradient is positive. As companies make progress through the execution of the orders, they tend to step up their competitive activities while still maintaining full focus on customer orientation to have a positive impact on performance. This trend is demonstrated by positive gradients for both customer and competitor orientations. It appears that companies do not lose focus of their customers at any time, but do adjust their level of competition-oriented activities given the resources they have to compete in the marketplace.

Another plausible reason could be that machine tool companies take a longer period of time to develop customer orientation than companies do in other industry sectors (cf. service industry). Therefore, by the time a firm is equipped to derive benefits from becoming competitor oriented, the nature of the competition may have changed drastically. Hence, it is important to assess the external environmental variables, such as technology change, market growth, etc. before a company attempts to commit resources to obtain or sustain such orientation. Certainly, companies may not be able to gain a competitive advantage quickly enough due to the need for massive investment in capital items, which may have an adversarial effect on the performance of companies.

The significance of the effect of competitor orientation upon performance calls for a better appreciation of the variables that influence the relationship. For example, while some companies may adopt a more competitor-oriented strategy, others may pursue cost or price-cutting measures to neutralise environmental pressures, such as market dynamism or differing levels of strength in the economy. Such benefits are expected to be short-term; however, in the long-term, this approach may have no effect on profitability (Appiah-Adu and Singh 1998). Therefore, the external emphasis may enable companies to find more opportunities in the market compared to their relatively less market oriented competitors.

Customer satisfaction focus: Customer satisfaction factor has a significant ($p < 0.00$) and a positive (0.49) effect on business performance. From Table 4.16 and Figure 4.2, it appears that medium and high customer satisfaction oriented companies tend to perform better than low customer satisfaction oriented companies. This finding is consistent with the conventional wisdom that customer orientation is likely to lead customer satisfaction, a factor, which has an influence on repeat purchase (Heskett and Jones 1994). In the machine tool industry, it is common that most of the sales volume is determined by repeat orders, and that these repeat orders are generated through satisfied customers. Companies cannot expect to obtain a good customer satisfaction rating by merely selling machines. In some cases, satisfaction can also be obtained by providing extra functionality to the existing machines. This is often done to increase operational efficiency, which may lead to customer satisfaction as well.

Further, customer satisfaction can be derived through the superior performance of machines and through the services a company can offer to its clients after sale (Singh and Ranchhod, 1998). Service after sales is an important source of revenue. Therefore, it is important that a machine has a good life span of operational capabilities coupled with quality service after sales, and that companies build customer loyalty as customers may seek services, upgrades or replacements for their machine tools. This kind of relationship, which may lead to customer loyalty, is particularly important for companies as customers can give a natural feedback to their manufacturers who, as a result, should strive to provide robust products that are able to perform under a variety of conditions. Research has shown that even if loyal customers buy competitor's product to take advantage of a special deal, they generally return to their original company for their next purchase (Deighton and Henderson, 1994). Also, loyal customers are more receptive to line extensions and other new products offered by the same company, and they are more likely to forgive an occasional product or service failure (Bejou and Palmer, 1998). However, what gives more cause for concern is that the majority of dissatisfied customers do not express their dissatisfaction with the performance of products or the delivery of services, they just move their custom elsewhere, destroying all the effort and investment put into improving customer satisfaction. On the other hand, customers who complain and receive a satisfactory response become more loyal to the company than those who have never complained because they now feel confident that the company would take extra care to resolve the problem. This implies that feedback from customers is vital, and that companies should use all their available tools, such as forms of feedbacks, reports of complains, findings of market research, etc. The premise is that customers should be encouraged to give feedback, via employee or toll-free number, that can be passed on to the concerned authority for corrective action as necessary.

Thus, as companies become customer- and competitor-focused and driven by customer demands, the need to meet the customers' expectations and retain their loyalty while maintaining long-term relationship becomes more critical. The results of the study suggest that high customer satisfaction leads to a high business performance, i.e. satisfied customers are much more profitable to companies than occasional buyers.

Marketing focus: This has a significant ($p < 0.00$) and a positive ($\beta = 0.24$) effect on business performance. Table 4.16 suggests that each group of marketing focus has a significant effect on business performance. From Figure 4.2, it appears that medium and high marketing-oriented companies perform better than low marketing oriented companies. Recently there has been a growing appreciation for the importance of marketing within this sector (*Financial Times*, March 18, 1996). The article explains how the company, 600 group, resorted to new marketing ideas which were once considered unusual in the conservative machine tool industry. To increase sales, the company screened for new clients by employing a team of telephone sales people from NWS Bank, a part of the Bank of Scotland. In exchange for this service, the bank gets an opportunity to propose a financial package should the deal materialise. The need for a marketing orientation is further confirmed by another article in the *Financial Times* (April 30, 1996) that explains that the machine tool sector has benefited from an improved marketing approach, which emphasises value for the money. Further, there is evidence that marketing has been a significant factor in the success of the Japanese and German companies (Doyle et al., 1985; Simon, 1991). Certainly, if marketing is an important determinant of success, it is worth paying attention.

Conclusions

The chapter reveals that there is a positive and significant relationship between market orientation and business performance. Performance was measured by return on investment, sales growth, market share, customer retention and global market presence of companies. Further examination of the relationship indicates that the market orientation is significantly and positively related to return on investment when market turbulence is high, to new product success when technological turbulence is low, and to sales growth when competitive intensity is high. Additionally, the study suggests that there is a significant positive relationship between market orientation and the overall performance of companies. The analysis of the data detects that the market orientation is significantly and positively related to overall performance when market and technological turbulences are low and competitive intensity is high.

Because the positive relationship between market orientation and performance depends on a particular performance indicator and the intensity of environment, it has implications for managers as to when a market-oriented strategy should be pursued to achieve defined objectives of companies, such as to increase return on investment, sales growth, market share, customer retention and global presence. If the objectives are not clear, a market-orientated strategy may not lead a company to be profitable. This is particularly true if companies operate in hostile conditions; however, the overall performance of companies can still be improved if the environment is hostile up to a certain level. Under such a situation, it is important to monitor the environment carefully in terms of the transition of the market and technological conditions. These transitions may change buyers' preferences. Further, there are conditions that require a high magnitude of market orientation, which may not be profitable for companies (Kohli and Jaworski, 1990).

Interestingly, it is worthy of note that there are four significant factors representing market orientation: customer focus, competitor focus, customer satisfaction focus and marketing focus. The influence of each focus on the performance suggests that customer and customer satisfaction focuses have stronger impacts on performance than the other focuses, and that competitor focus has a U-shape relationship with performance. Managers could use the multidimensional conceptualisation to develop particular kinds of focus for their companies to enhance business performance by responding (thus obtaining a composite market orientation score) to the items on the market orientation scale at any point in time. Then a target should be set to achieve a higher score within a time frame on those items that needed improvement. The target can also be considered as a benchmark, which can be set by forming focus groups within the organization whose members are drawn from various departments, as one of the goals of being market-orientated is to have interfunctional co-ordination among departments. A repetitive use of the scale in future to measure the differences between the scores on the items and the scores on the bench-marked items will determine the extent to which managers are successful in achieving market orientation. Attention must be paid to the individual items (statements) of the scale when calculating and interpreting the market orientation score between two points in time, because many patterns of response to the various items can produce the same total score. Therefore, identical total scores may reflect different attitudes because of different combinations of items endorsed.

In the next chapter, I add another perspective to market orientation - corporate culture. The impact of the corporate culture on market orientation is examined.

Chapter 5

Corporate Culture and the Market Orientation

Introduction

This chapter focuses on the corporate culture as an antecedent to the market orientation. Further, it explores the linkage between the dimensions of the corporate culture and the market orientation. In the remainder of this chapter, a conceptual background on market culture in the context of corporate culture is presented, leading to testing a hypothesis. The hypothesis states that the degree of market orientation varies respectively from highest to lowest according to the type of corporate culture as follows: market, adhocracy, clan, and hierarchical culture. Findings suggested that the degree of market orientation of firms operating in the machine tool industry in the UK varied respectively from highest to lowest according to the following type of corporate culture: adhocracy, market, hierarchical and clan culture. The difference in hypothesis and the findings is explained in the discussion section of the chapter.

Significance of the Culture

Notwithstanding the scant empirical research on the culture, the subject has been the focus of debate and conceptual research for several years. Culture has been discussed extensively in the academic literature and many definitions and conceptualisations of the culture have been proposed (Peters and Waterman, 1982; Schein, 1985). Some definitions provided in the pertinent literature include: a pattern of beliefs and expectations shared by organization members (Schwartz and Davis, 1981), and some underlying meaning that persists over time, constraining people's perception, interpretation and behaviour (Jelinek et al., 1983). Schneider and Rentsch (1988) described the culture as the unwritten policies and guidelines (the way we do things here) which are believed to influence behaviour and performance of employees more than the formal procedures and systems. Culture is a critical component which executives might utilize to shape the direction of their businesses (Smircich, 1983), particularly because culture affects productivity as businesses adapt to business environment, and to the new employees (Schneider and Arnon, 1983). These pervasive influences reinforce the need for organizations to pay close attention to the concept of culture in their operating environments.

Due to the recognition of the role played by an organization's culture in the management of marketing function, a related concept, which has attracted growing interests among scholars and managers, is how organizations can cultivate a market culture. Because a market culture regards the whole business from a customer's perspective, the philosophy represents a corporate culture - a basic set of values and beliefs - which places the customer at the heart of the organization's approach to strategies and operations (Deshpande and Webster, 1989). Marketing scholars have investigated the influence of culture on a variety of management and marketing-related variables; for example, culture has been proposed as a key determinant of success in implementing marketing strategies (Walker and Ruekert, 1987), influencing sales (Weitz et al., 1986), and enhancing management effectiveness (Parasuraman and Deshpande, 1984).

The Definition and the Corporate Culture

Corporate culture is defined as the pattern of shared values and beliefs that help individuals understand organizational functioning, and thus provides norms for behaviour in the organization (Deshpande et al., 1993). A great deal of attention has recently been devoted to the market-oriented culture that is necessary to facilitate the achievement of an organization's goals. In the context of market orientation literature, emphasis is placed upon the deployment of an organization-wide effort to satisfy customer needs (Kohli and Jaworski, 1990; Narver and Slater, 1990; Jaworski and Kohli, 1993; Day, 1994; Slater and Narver, 1995). For an effective market orientation, these studies emphasize a corporate culture that delivers superior value to customers via a coordinated cross-functional effort. Therefore, the importance of corporate culture is rooted in the fact that it can serve as a tool to improve productivity and if properly communicated, it can be used to encourage all employees to subscribe to the organizational goals (Deal and Kennedy, 1982; Wilkins and Ouchi, 1983; Block, 1989; Stewart, 1990; Alan, 1998).

Drawing on Smircich's (1983) conception, Deshpande and Webster (1989) discussed five paradigms for an organizational culture. Of which organizational cognition is the most developed perspective on an organizational culture. This is based on management information processing and perceives organizations as knowledge-based systems. Webster and Deshpande (1990) discuss the significance of this information-processing viewpoint in the understanding of organizational culture by drawing upon Quinn and Rohrbaugh's (1983) concept of competing values within organizations. Table 5.1 presents various organizational paradigms.

Market Orientation as a Culture

The market orientation concept has basically two perspectives: behavioural (Kohli and Jaworski, 1990; Deng and Dart, 1994; Doyle and Wong, 1998; Jaworski et al., 2000); and cultural (Slater and Narver, 1995; Turner and Spencer, 1997; Harris, 1998; Narver et al., 1998). From the cultural perspective, market orientation is perceived as a corporate culture which possesses a set of shared values and attitudes throughout the organization that guides individuals' behaviour (Linchtenthal and Wilson, 1992) in a way that stimulates creation of values for customers (Pearce and David, 1987; Greenley, 1995) which over a period of time becomes the focal point of strategy and organizational actions (Deshpande and Webster, 1989). Consistent with this perspective, Norburn et al. (1990) contend that a market-oriented business should have a close relationship with customers and an external orientation to its markets. Customer relationship is characterized as a service orientation, a drive towards innovation and a focus on quality from the customer's viewpoint. The external focus on markets recognizes the significance of the marketplace as an important influence on corporate action.

In an effort to determine the factors that distinguished superior organizations from their competitors, Peters and Waterman (1982) identified two important

Table 5.1 Organizational paradigms

Organizational Paradigm	Key Theoretical Features	Locus of Focus
1 Comparative management	Grounded in functionalism (Malinowski, 1961) and classical management theory (Banard, 1938)	Exogenous, independent variables
2 Contingency management	Grounded in structural functionalism (Radcliffe-Brown, 1952) and contingency theory (Thompson, 1967)	Endogenous, independent variables
3 Organizational cognition	Grounded in ethnoscience (Goodenough, 1971) and cognitive organizational theory (Weick, 1979)	Culture as metaphor for organizational knowledge systems
4 Organizational symbolism	Grounded in symbolic anthropology (Geertz, 1973) and symbolic organization theory (Dandrige et. al., 1980)	Culture as metaphor for shared symbols and meanings
5 Structural/ psychodynamic perpective	Grounded in structuralism (Levi-Strauss, 1963) and transformational organizational theory (Turner, 1983)	Culture as a metaphor for unconscious

Source: Adapted from Smircich (1983) and Deshpande et al. (1989)

organizational factors: effective marketing practices and customer orientation. Hooley and Lynch (1985) found that a common characteristic of the best performing UK firms was marketing excellence. Kiel et al. (1986) made similar observations in their study of Australian firms. In Japan, economic performance was attributed to the focus on marketing for global sales (NEDO, 1982). Webster's (1988) findings suggest that marketing is gaining increasing attention among executives due to its importance as a competitive tool in the business environment. It is contended that effective marketing will not only distinguish the amateur from the professional players in the global market, but also it has a potential to launch a new era of economic prosperity and high living standards (Ghosh et al., 1994; Kotler, 1996).

Therefore, it is necessary to recognize the importance of studying the market, identifying the numerous opportunities, selecting the best segments of the market, and making an effort to provide a superior value to satisfy the customers' needs. In addition, the firm must be adequately staffed to generate information, disseminate that information within the organization and take actions with the view to satisfy customers' needs (Kohli and Jaworski, 1990). Successful implementation of market orientation demands that adequate information is available to managers for the purposes of planning and allocating resources to markets, products and territories.

Several scholars describe the marketing concept (implementation of the concept is referred as market orientation) as a form of corporate culture. For example, according to Deshpande and Webster (1989), the marketing concept is a distinct organizational culture, a fundamental set of shared beliefs and values that puts the customer in the centre of the firm's thinking about strategy and operations. Narver and Slater (1990) stress this viewpoint by asserting that it is the organizational culture that most effectively creates the necessary behaviour for creating a superior performance for the business. Kohli and Jaworski (1990) use the term market orientation to describe organizational behaviour and activities that manifest the adoption of the marketing concept. In keeping with the preceding definition, market orientation is used in this chapter as an organizational behaviour at firm level. I use organizational culture and corporate culture interchangeably.

Market orientation is considered as a cultural orientation, thus, an element of the organization, which is expected to affect a firm's learning capability and its strategic resource position. Market orientation reflects a culture that encourages organizational learning behaviour to create and maintain profitable relationships with customers (Slater and Narver 1995). Further, Day (1994) suggests that market-driven organizations have superior market sensing, customer linking and channel bonding capabilities. Therefore, market orientation, when discussed in a cultural sense, focuses more upon corporate culture. Logically, previous arguments tend to support the notion that market orientation can be considered to be positively associated with organizational culture which in turn may lead organizations to perform better.

Market Orientation and the Culture Types

Despite the widespread acceptance of the marketing concept in principle, the development of management skills required to effect marketing plans still remains a problem area for many firms. Research findings suggest that only a few businesses really understand and perform sophisticated marketing activities. For instance, Hooley and Newcomb (1983) observed that British firms were characterized by dominant production-oriented culture and a general lack of market-oriented behaviour. Doyle et al. (1985) showed that in comparison with their Japanese counterparts, British firms tended to be more financial or production oriented than market-focused.

In the US, Kotler (1977) observed that marketing was one of the most misunderstood organizational functions and that only a few Fortune 500 firms applied sophisticated marketing practices. Later surveys suggested that the cultivation of a marketing focus was still a problematic area within many organizations, with a large number performing below their full potential (Webster, 1981; Payne, 1988).

One of the reasons why firms fail to achieve marketing success is the existence of organizational barriers which hinder the implementation of marketing programs and customer-focused strategies. These obstacles include both organizational and marketing functions, and their related systems and policies (Webster, 1988). However, beyond this argument is the contention that there is a more pertinent underlying issue associated with the poor performance of marketing activities by firms, such as the human element involved in developing and implementing marketing strategies or the organization's market culture (Dunn et al., 1994).

Although management and marketing literature is filled with discussions on various types of cultures, such as market culture, corporate culture, and national culture, market culture represents a latent culture, which is often not formally decreed. Quinn (1988), and Cameron and Freeman (1991) developed a typology for the types of corporate culture, which is presented in Figure 5.1

The one end of the horizontal axis represents internal maintenance (integration and smoothing operations) and the other end external positioning (environmental differentiation, competition) whereas the one end of the vertical axis represents processes (stability, order, control) and the other end organic (spontaneity, flexibility, individuality). This conceptualisation results in four main types of culture - adhocracy, clan, hierarchical and market culture. These culture types are the dominant ones rather than mutually exclusive classifications. Although the majority of organizations may be characterized by more than one form of culture, one particular type of culture assumes a predominant position over a period of time (Deshpande and Webster, 1989).

Deshpande et al. (1993) tested for differences in business performance in a sample of Japanese firms on the basis of these four culture topologies, and found that performance ranked from best to worst according to the type of corporate culture. The best performance was observed from the companies which had a market culture, followed by an adhocracy culture, a clan culture and a hierarchical culture. The adhocracy culture, with its emphasis on entrepreneurship, innovation, and risk taking ability is expected to have a relatively higher degree of market orientation than the clan culture. The clan culture, which stresses tradition, loyalty, and internal maintenance could result in a lack of attention to changing market needs, leading to a low degree of market orientation. The hierarchical culture, with its focus on smooth operations and predictability in a bureaucratic organization is likely to lead to a low

ORGANIC PROCESS (flexibility, spontaneity)

TYPE: Clan	TYPE: Adhocracy
DOMINANT ATTRIBUTES: Cohesiveness, participation teamwork, sense of family	DOMINANT ATTRIBUTES: Entrepreneurship, creativity adaptability
LEADER STYLE: Mentor, facilitator, parent-figure	LEADER STYLE: Entrepreneur innovator, risk-taker
BONDING: Loyalty, tradition, interpersonal cohesion	BONDING: Entrepreneurship, flexibility, risk
STRATEGIC EMPHASIS: Toward developing human resources, commitment, morale	STRATEGIC EMPHASIS: Toward innovation, growth, new resources
INTERNAL MAINTENANCE (Smoothing activities, integration)	EXTERNAL POSITIONING (competition, differentiation)
TYPE: Hierarchical	Type: Market
DOMINANT ATTRIBUTES: Order, rules and regulations, uniformity	DOMINANT ATTRIBUTES: Competitiveness, goal achievement
LEADER STYLE: Coordinator, administrator	LEADER STYLE: Decisive, achievement-oriented
BONDING: Rules, policies and Procedures	BONDING: Goal orientation, production, competition
STRATEGIC EMPHASIS: Toward stability, predictability, Smooth operations	STRATEGIC EMPHASIS: Toward competitive advantage and market superiority

MECHANISTIC PROCESSES (control, order, stability)

Figure 5.1 A model of organizational culture types

Source: Adapted from Quinn (1988) and Cameron and Freeman (1991)

level of market orientation. On the other hand, the market culture, which is based on differentiation, competitive advantage, and market superiority, is expected to exhibit a high level of market orientation. Hence, organizational emphasis on external positioning over internal maintenance is likely to be associated with higher levels of market orientation. Therefore, drawing upon the above notion, the hypothesis to be tested is as follows:

> *H3:* The degree of market orientation varies respectively from highest to lowest according to the type of corporate culture as follows: market, adhocracy, clan, and hierarchical.

The graphical representation of the hypothesis is depicted in Figure 5.2.

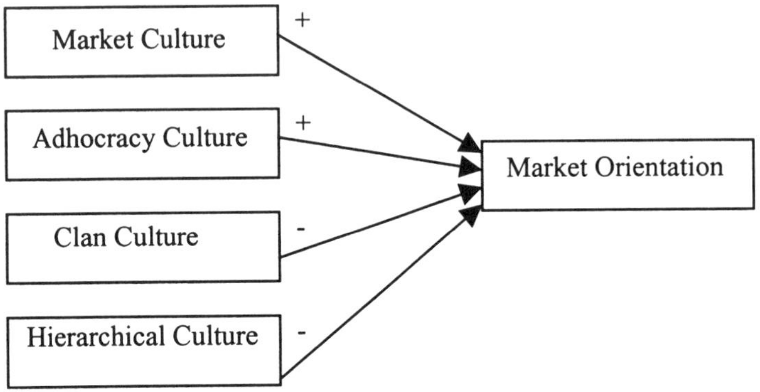

Figure 5.2 Antecedents to the market orientation

Source: Singh (1998)

The Research Methodology

The Measures

The following constructs were used to measure the corporate culture and the market orientation variables.

Corporate culture: The construct was adapted from Deshpande et al. (1993), which was drawn upon the studies of Quinn (1988), and Cameron and Freeman (1991). The corporate culture has four dimensions: market culture, adhocracy culture, clan culture, and hierarchical culture. Each dimension has four items representing each distinct form of corporate culture. For example, market culture is a proxy for behaviours, such as achievement and goal orientation, working towards maintaining competitive advantage and market superiority, whereas adhocracy culture is represented by the need for innovativeness, entrepreneurship, and risk taking abilities of firms. Similarly, clan culture stresses loyalty, tradition, and emphasis on internal maintenance. Finally, hierarchical culture places emphasis on predictability and smooth operations in an organization.

In this study, a constant sum scale was used for the measurement of corporate culture. Respondents were asked to distribute 100 points among the four descriptions depending upon how they perceived these statements to be reflective of their businesses. This methodology requires respondents to allocate a fixed number (usually 100) among several objects to reflect the relative preference for each object (Guilford, 1954). It is widely used to measure the relative importance of attributes. Therefore, this methodology was employed to analyze the data in this chapter. A questionnaire containing these items is presented in Table 5.2.

Table 5.2 Corporate culture

These questions relate to what your operation is like. Each of these items contain four descriptions of organizations. Please distribute 100 points among the four descriptions depending on how similar the description is to your business. None of the descriptions is any better than any other; they are just different. For each question, please use all 100 points. You may divide the points in any way you wish. Most businesses will be some mixture of those described.

1. Kind of organization (Please distribute 100 points)

A. My organization is a very personal place. It is like an extended family. People seem to share a lot of themselves.

B. My organization is a very dynamic and entrepreneurial place. People are willing to stick their necks out and take risks.

C. My organization is a very formalised and structured place. Established procedures generally govern what people do.

D. My organization is very production oriented. A major concern is with getting the job done without much personal involvement.

2. Leadership (Please distribute 100 points)

A. The head of my organization is generally considered to be a mentor, sage, or a father or mother figure.

B. The head of my organization is generally considered to be an entrepreneur, an innovator, or a risk taker.

C. The head of my organization is generally considered to be a co-ordinator, an organiser, or an administrator.

D. The head of my organization is generally considered to be a producer, a technician, or a hard-driver.

3. What holds the organization together (Please distribute 100 points)

A. The glue that holds my organization together is loyalty and tradition.

Commitment to this firm runs high.

B. The glue that holds my organization together is a commitment to innovation and development. There is an emphasis on being first.

C. The glue that holds my organization together is formal rules and policies. Maintaining a smooth-running institution is important here.

D. The glue that holds my organization together is the emphasis on tasks and goal accomplishment. A production orientation is commonly shared.

4. What is important (Please distribute 100 points)

A. My organization emphasizes human resources. High cohesion and morale in the firm are important.

B. My organization emphasizes growth and acquiring new resources. Readiness to meet new challenges is important.

C. My organization emphasizes performance and stability. Efficiency, smooth operations are important.

D. My organization emphasizes competitive actions and achievement. Measurable goals are important.

Source: Quinn (1988) and Cameron and Freeman (1991)

Market Orientation: The market orientation score was split at median into two groups, the median being 4.7 on a seven-point scale. The groups with a value higher than the median were referred to as high market-oriented groups whereas groups lower than the median value were referred to as low market-oriented groups.

Reliability and Validity

Reliability: Table 5.3 reports the scores for alpha and standardised alpha for each dimension of the corporate culture. Reliability for the market orientation scale has been reported in Tables 3.17a and 3.17b in Chapter 3. Table 5.3 suggests that the scores for the dimensions of corporate culture are within the acceptable range and greater than the suggested cut-off level of 0.60 (Nunnally, 1955) except hierarchical culture. Though hierarchical culture has a relatively low reliability coefficient, it was retained in the analysis for theoretical purposes because it is part of the broader conceptual framework (Quinn, 1988; Cameron and Freeman, 1991). It may be observed from Table 5.3 that there is a little difference between the scores of alpha and the standardised alpha (this compensates for the effects of the number of scale items), thus, lending credence to the reliability of market culture, adhocracy culture, and clan culture constructs. These Cronbach alpha scores for the dimensions of corporate culture reinforce the reliability of the internal consistency of their respective items. The reliability analysis for the hierarchical culture suggests that there is a low correlation (0.07) between policies and stability items. It appears that in the context of the machine tool industry, stability, predictability or smooth operation contribute little to hierarchical culture. Even after excluding the stability item, the reliability for the hierarchical dimension could not be increased.

Table 5.3 Reliability analysis for the corporate culture

Cultural Dimension	No of items	Cronbach α	Std. Cronbach α
Market Culture	4	0.63	0.64
Adhocracy Culture	4	0.60	0.61
Clan culture	4	0.70	0.71
Hierarchical	4	0.36	0.40

Content validity: It is reasonable to believe that the corporate culture has content validity as it has been grounded in theory and tested previously (Narver and Slater, 1990; Jaworski and Kohli, 1993; Deng and Dart, 1994).

Discriminant Analysis

Discriminant analysis is appropriate when the dependent variable is categorical (the market orientation variable was split into two categorical groups, low and high market-oriented groups) and the independent variables are metric (Norusis, 1993). The analysis involves deriving the linear combination of the four independent variables (each of the four dimensions of the corporate culture was taken as an independent variable) that will discriminate best between the low and high market-oriented groups. This is achieved by the statistical decision rule of maximising the between-group variance relative to the within-group variance. This relationship is expressed as the ratio of between group to within group variance. The linear combinations for a discriminant analysis are derived from the following equation.

$$Z = W_1 X_1 + W_2 X_2 + W_3 X_3 + \ldots W_n X_n$$

Where, Z=the discriminant score, W=the discriminant weight, X=the independent variable.

This analysis is a suitable statistical technique for testing the hypothesis where the means of two or more groups are equal. In this technique, each input variable is multiplied by its corresponding weight and then the products are added together. The result is a single composite discriminant score for each case in the analysis. By averaging the discriminant scores for all cases within a particular group, I can arrive at the group mean. This group mean is referred to as a centroid. Each group has a centroid, which indicates the most typical location of a case from a particular group. Comparison of these group centroids show how far apart the groups are along the dimensions being tested.

The test for the statistical significance of the discriminant function is a generalised measure of the distance between the group centroids. It is computed by comparing the distribution of the discriminant scores for the low and high market-oriented groups. If the overlap in the distribution is small, the discriminant function separates the groups well. If the overlap is large, the function is a poor discriminator between groups.

All dimensions of corporate culture were entered into the model, as the objective of the analysis was to determine simultaneously the discriminatory capabilities of the four dimensions. Wilks' lambda provides the discriminatory significance for each dimension of the corporate culture. Discriminatory power of the model is calculated in the following section.

Discriminatory Power of the Model

The classification results are reported in Table 5.4. The discriminant function indicated that 59.14 per cent of the cases were correctly classified.

Table 5.4 Classification results

	Predicted group		
Actual Group (Actual total)	LO-MO firm	HI-MO firms	Total firms
Low Market Oriented Firms	25 (58.10%)	18 (41.90%)	43
High Market Oriented Firms	20 (40.00%)	30 (60.00%)	50
Predicted Total	45	48	93

C_{model} (59.14%) > $C_{maximum}$ (53.76%) > $C_{proportional}$ (50.17%)

To find out the discriminatory power of the model, it was intended to compare the model with the priori chance of classifying cases correctly without the discriminant function. As I have unequal group sizes (the low market-oriented group has 43 cases whereas the high market-oriented group has 50 cases), the following formula was used for the calculation of proportaional chance criterion.

$$C_{proportional} = p^2 + (1 - p)^2$$

Where, $C_{proportional}$=the proportional chance criterion, p=proportion of cases in group 1, and (1 - p)=proportion of cases in group 2.

Substituting the appropriate numbers in the formula, I obtain:

$$C_{proportional} = (43/93)^2 + (1 - 43/93)^2 = 0.5017 = 50.17\%$$

The proportional chance criterion (50.17 per cent), which is lower than 59.14 per cent, indicates that the discrimant function has more discriminatory power than the probability calculated using proportional criterion.

Further, the maximum chance criterion was calculated to find out the percentage of correctly classified cases if all the cases were placed in one group with the greatest possibility of occurrence. Since the high market-oriented group occurs $(50/93)^2$ = 53.76 per cent of time, it would be correct 53.76 per cent of time if all cases were assigned to this group. Because the value for the maximum chance criterion is larger than the value for the proportional chance criterion, the discriminant model should outperform at 53.76 per cent of time. Therefore, it can be concluded that the discriminant model has the highest discriminatory power (Morrison, 1969).

Results

The mean scores and pooled correlation coefficients are reported in Table 5.5 which suggests that adhocracy culture (0.57, $p < 0.05$) is the most significant determinant of market-oriented companies followed by, in sequence, market culture (0.44), hierarchical culture (-0.33) and clan culture (-0.21). The coefficients are not significant for hierarchical ($p > 0.05$) and clan culture ($p > 0.05$). Coefficients of these four culture types do not follow exactly as expected but they do resemble the same pattern of culture types. For example, there is little distinction between market culture and adhocracy culture for a company to be market-oriented. Although not significant, clan and hierarchical culture appear to be negatively related to market-oriented companies.

Table 5.5 Discriminant analysis

Dimensions of corporate Culture[a]	Means[b]			standard deviation	structured matrix corr.[c]	F ratio[d]	sig.[e]
	All	LO-MO	HI-MO				
Adhocracy	24.92	22.71	26.82	9.81	0.57	4.18	0.04
Market	22.26	20.51	23.92	9.73	0.44	3.48	0.06
Hierarchical	20.59	21.45	19.85	8.36	-0.33	0.85	0.35
Clan	33.10	33.83	32.47	14.19	-0.21	0.20	0.64

[a] Culture dimensions are rank ordered according to size of correlation with the function.
[b] Numbers are mean of the four individual components for each culture type.
[c] Pooled within-groups correlations of discriminant function and independent variables.
[d] Significant univariate difference between low and high market oriented firms (with 1 and 91 degree of freedom).
[e] Significance for F univariate ratio.

Discussion

Our findings reveal that adhocracy culture has a significant and positive (0.57) effect on market-oriented companies. It is followed by a market culture, which is almost significantly related to market-oriented companies. It appears that both adhocracy and market culture appear to be the most requisite cultures for high market-oriented companies. The adhocracy culture seems to be the dominating culture because it emphasizes entrepreneurship, innovativeness, flexibility, and risk-taking ability of firms.

In the context of the machine tool industry, this is what would be expected because a strong desire for independence and a fascination with technology leads a company to form strong emotional ties between entrepreneurs and the products which over a period of time shapes the culture of the organization. Being most of the machine tool companies small- and medium-sized, one of the ways to be competitive is to have innovation orientation. Mr. Jean-Pierre Lefebvre, president of LVD machine tool Company, who places significant importance on adhocracy culture, suggests that the best way to come out of recession is to innovate, particularly to survive a recession climate (*Financial Times*, 1994). Firms with entrepreneurial proclivity are innovative, and a central theme of the innovation literature is that information gathering and analysis are critical to the successful development and execution of innovation strategies (Barringer and Bluedorn, 1999; Covin, 1991). Furthermore, entrepreneurial firms tend to engage in a greater level of information-scanning activities (Hambrick, 1982). The process of information gathering itself is a substantial part of being market oriented.

Flexibility within companies associated with adhocracy culture seems to be another determinant of high market-oriented companies. Because the SMEs are less formal and centralised, the adhocracy culture appears to lead to speedier decisions and actions which are more likely to be adapted to individual situations. Lumpkin and Gregory (1996) echo the view by arguing that entrepreneurial-oriented companies value autonomy and freedom to encourage creativity and champion untested but promising ideas. Thus it appears that machine tool firms are determined to maintain adhocracy culture to foster creativity, entrepreneurship and ability to take risk that has contributed positively to market orientation.

Next in the sequence is the market culture, which is positively (0.44) related to market-oriented companies. This finding is consistent with our expectation that market culture is positively related to market orientation though it appeared as the second most dominant culture. The characteristics of this culture include competitiveness, goal achievement, productivity, competitive advantage and market superiority. This culture promotes behavioural norms required for a firm to develop and respond to market information to create a sustainable competitive advantage over their competitors. This particular culture has been adopted by Europe's largest lathe machine manufacturer, the 600 Group, which has developed a lathe at a 30 per cent lower cost than an old machine launched a few years ago with a similar set of specifications by deleting the features that did not frequently contribute to productivity. It was the outcome of an extensive market research which indicated that the customers wanted more value for the money. As a result, the company decided to design a machine whose emphasis was on value addition and on market orientation. Measures were taken to bring the cost of the machine down by removing undesirable features of the machines, e.g. third axis, live tooling, etc. The new lathe machine was a huge success, which significantly boosted the market share of the company worldwide. The culture of the company is driven by the market rather than its endeavour to sell what it could produce. In Japan, Deshpande et al. (1993) found that firms with a strong market culture demonstrated a higher level of business performance than their counterparts with a relatively weak market culture. Market culture is largely instituted by corporate individuals in the organization, which calls for employees to be united by a common set of corporate beliefs and values. Through these norms, a suitable environment is created for management to demonstrate a commitment to personal empathy, to promote a customer response of perceived quality, and ultimately, develop a progressive organization (Norburn et al., 1988). Certainly, we should not be surprised if the global universality of market culture might transcend into a national culture.

The Hierarchical culture is negatively (-0.33) related to market-oriented companies. In line with our expectations, the hierarchical organizations were found to be least market-oriented. This culture is characterized by placing emphasis on internal maintenance and smooth operation of firms, i.e. the result is an inward-looking bureaucratic firm, which is not customer- or market-oriented. Attributes associated with this culture are: formulation of rules, regulations, policies and procedures. The interview with a product strategy and marketing director of the 600 Group Company revealed that for thc company to be market-oriented, a degree of hierarchical culture was also required. It is important to formulate necessary rules, regulations and policies, which are just enough to keep the company's operations running smoothly without losing its market-orientated focus. Indeed, hierarchical culture gives rise to stability, uniformity, predictability and smooth administration. Although many businesses have structures, procedures and systems to guide them through marketing operations, the human dimension has a significant influence on successful execution of marketing plans, because their formulation and implementation depend on individuals in the firm (Piercy and Morgan, 1994). While it is true that greater formalization and centralization produce uniformity of policy and action, lessen risks of errors by personnel who lack either information or skill, utilize the skills of central and specialized experts, and enable closer control of operation (Flippo, 1966), it is also correct that excessive of any of these characteristics will lessen the attainment of market orientation. High hierarchical culture may mean more formalization, centralization, or departmentlization in addition to more rules, regulations, procedures, departments, etc. which in turn would decrease the activities relating to intelligence generation, dissemination and responsiveness; thus, less attainment of market orientation. Kohli and Jaworski (1990) found that the centralization was negatively related to all three dimensions of the market orientation. Therefore, appropriate culture required for a market orientation is the one that is built upon a customer orientation and that permeates through an entire organization without being obstructed by extra rules and regulations. Development of a suitable culture, without being too hierarchical, is crucial for the implementation of a market orientation. It appears that machine tool managers understand the trade off between market orientation and hierarchical culture; indeed a balance is necessary for companies' optimum performance.

Finally, the clan culture is negatively (-0.21) related to market-oriented companies. This culture appears to be the least desirable culture by firms for the enhancement of their market orientation. The clan culture considers the entire organization as a family and a godfather is always responsible for all the activities concerned. Although teamwork, participation and loyalty are part of the culture, it seems to be less prevalent. With reference to the machine tool industry, it appears that companies perceive factors like delegation, taking responsibility and empowerment to be more important than depending upon a godfather. It may be suggested that machine tool companies practise professionalism more than loyalty or interpersonal cohesion. Further, it seems that interpersonal skills may not be a necessary determinant of high market-oriented companies, as employees may have a little opportunity to interact with the market. In particular, Japanese firms, though tending to be clan, also show a strong element of market culture (Deshpande et al., 1993). However, in the setting of the UK, I found the clan culture to be less desirable than the rest of the culture types. Differences in the findings, based on the sequences, were found because of the two different international settings, namely the UK and Japan. These differences are consistent with the findings from Sullivan (1983) and Hatvany and Pucik (1981) as they noted a considerable diversity in both structural and cultural Japanese organizations which were seldom mentioned in academic writing. Another plausible reason for the differences in the sequence of the cultures could be the fact that some firms exhibited a variety of culture types.

Conclusions

This chapter focused upon the potential linkages between corporate culture and market orientation. Although, the chapter adopts the corporate culture measure from Deshpande's (1993) study, additional issues relating to market culture and market orientation were examined. Specifically, this study is extended to examine whether the corporate culture is an antecedent to market orientation of firms, and whether there is a positive relationship between corporate culture and market orientation in the machine tool industry.

A forging of links between market culture and market orientation appears to be an important antecedent to success. In accordance with the argument that the market culture is an important determinant of an organization's superior performance, the evidence found in this chapter confirms that there is a positive association between adhocracy culture, market culture and a high level of market orientation.

Our findings suggest that the degree of market orientation of firms operating in the machine tool industry varies respectively from highest to lowest according to the following type of corporate culture: adhocracy, market, hierarchical and clan culture. Although results suggest some indication as to the ranking of the corporate culture, they do not support the hypothesis. Market culture was hypothesised to be the most significant culture followed by adhocracy culture. Both cultures were expected to be significantly and positively related to market-oriented firms. The results showed that the adhocracy culture was the most significant culture followed by the market culture for the market-oriented companies. Thus there was a difference in the sequence of the cultures. The next dominant hypothesised culture was clan culture followed by hierarchical culture. These cultures were expected to be negatively related to market orientation of companies. Findings indicated that the next dominant culture was hierarchical followed by clan culture. As expected, I found that both cultures were negatively related to market-oriented companies. Because the means of all four cultures were well represented in the sample, findings suggest the extent to which firms have adopted the combination of cultures rather than any pre-dominant culture. The positive link observed between adhocracy culture, market culture and market orientation implies that in order to achieve high level of market orientation managers need to have an understanding of the impact of cultural issues.

Our finding that high level of adhocracy culture is positively related to high level to market orientation leads managers to face an important issue in understanding how to promote innovative entrepreneurial proclivity in their organizations. For example, top managements' expressed commitment to entrepreneurial risk-taking and its continuous risk-taking initiatives across different departments seem to be helpful. For most high-tech companies, innovation is their lifeblood. Yesterday's innovations do not necessarily satisfy today's customers. So, managers of successful companies not only need innovation but also a repeatable process to drive a continual stream of innovative products. By taking a new approach to solving problems, managers can dramatically re-define an existing market.

Next, a market culture is equally important for firms inspiring to achieve superior performance in their competitive environments. Managers can instil the culture in their employees by emphasising on achievement-oriented leader style, taking competitive advantage and market superiority, and forming a bond that is based on goal and competition orientations. The hiring of new employees with a strong market focus may assist in expediting the rate of learning through training, which in turn may serve as a vehicle to sustain market-oriented culture. Relevant training and professional development can help employees enhance their market-focused skills and increase their value to the department. These initiatives are consistent with the goal of increasing market orientation that requires organization-wide, continuous learning about the market (Jaworski and Kohli, 1993). Additionally, when recognizing employees who have achieved market-oriented results, managers should endeavour to be as equitable as possible. Further, it is important to remember employees who are yet to be market-focused but are making significant progress toward their performance goal.

With respect to hierarchical culture, managers should keep the level of rules and regulation as low as possible without sacrificing the need for stability, predictability and uniformity. Managers should define these policies for their employees in a way departments can best support the corporate strategic mission of the firm and clarify for each individual their role in helping the department achieve these objectives. Support for the notion also come from Narver and Slater (1990) who state that interfunctional coordination among department is vital for the achievement of market orientation.

Our findings do not lend support for the clan culture as it is detrimental to achieving market orientation. Therefore, managers perhaps should not dwell excessively upon the long tradition of doing things, fostering a sense of family or mentoring as a father figure in their organizations. However, these characteristics may influence market orientation if firms are very small or are usually run by only family members and/or distant relatives. In this sense, our finding is inconclusive; therefore more research is required to develop insight as to under which circumstances the relationship between clan culture and market orientation is positive. The next chapter presents a brief overview of the role of the British machine tool industry with its trends and competitiveness.

Chapter 6

The Machine Tool Industry

Introduction

The chapter offers a general overview of the machine tools and its definition followed by the function and evolution of machine tools. Next, a brief description of the machine tool industry is presented by providing relevant data which form a basis for discussion of case studies presented in Chapter 8. Finally, the chapter concludes with recent trends in the machine tool technology.

The Machine Tools

Machine tools are the most important means of producing large numbers of identical components. Without the development of machine tools, the high living standard of the present time would be unthinkable. In some of the most highly industrialised nations, approximately ten per cent of all the machines built are machine tools, and about ten per cent of the workforce in machine manufacturing is concerned with machine tools.

The technology of the machine tool industry is vital for the creation of all modern products. Machine tools are used by all sectors of manufacturing including aerospace, automotive, defence, railways, construction equipment, agricultural machinery, consumer durable and many other applications. Because machine tool manufacturers typically sell their products worldwide, a weak domestic machine-tool industry means that manufacturers risk losing access to the foreign markets. Therefore, the machine tool industry is fundamental to a nation's economy.

The Definition of Machine Tool

According to MTTA, a machine tool is defined as a power driven machine, not portable by hand when in operation, which works metal by cutting, forming, physiochemical or non-contact machining, e.g. lasers or electrical erosion machines, or a combination of these techniques. Metal cutting machines include lathes, grinding machines, milling machines, and machining centres. Typical metal forming machines are presses, forges, and punching, shearing, and bending machines. This product classification conforms to the Standard Industrial Classification Codes: 3541, metal-cutting; and 3542, metal-forming. Machine tools are designed for cutting, bending and shaping pieces of cast or forged metals. The part that does the work of the machine is called the tool, and the part being worked on is called the work piece.

Functions of the Machine Tools

There are a variety of operations that a machine tool performs. For example, turning, to reduce the diameter of a bar; drilling, to make holes; boring, to enlarge the diameter; tapping, to cut screw threads inside holes; milling, to shave material off flat surfaces; grinding, to produce a fine finish; shaping, to scrape the surface off a metal plate; shearing, to cut metal plate into lengths; profiling, to profile the work piece; and others.

Evolution of the Machine Tools

Operating a conventional machine tool is a highly skilled task. The operator positions the work piece, selects the appropriate tools in their correct sequences for each job, and manually operates the handles and levers that control the relative position and speed of the tool and work piece to produce parts with the required dimensions and features.

The degree of automation of machine tools is as varied as its area of application. One of the major areas of the operation of machine tools is cutting. Innovation in the machine tool design is only limited by the degree of technological development. Automation and flexibility of machine depends on the nature of the component and the quantities to be produced. For example, single-purpose or special-purpose machines are available to the user, as are universal machines offering a wide range of applications.

The increased demands in both performance and the precision of machine tools saw further automation of machine tools. This automation has resulted in the development of a wide range of alternative controls. In recent years, the development of electronics and computer programs has a marked effect on machine tool controls. The arrival of the microprocessor chip has made control techniques far easier than before. In the next sections, I describe broadly the phases of machine tool development.

Towards the end of the nineteenth century, it became possible to make machine tools carry out a sequence of operations automatically by employing complex arrangements of cams and gears. It was therefore no longer necessary to have one operator at each machine. The operator's job was reduced to loading the work piece, switching on the machine, removing the finished part, and replacing dull tools. Removal of dull tools and resetting the machine to carry different sequences of operations was a skilled and time-consuming job. Although these machines could be altered to carry out a wide range of different functions, the long set up times and high costs of new hardware programmes meant that they were only effective with comparatively long production runs or with job order applications where each job required different kinds of setting of tools.

In the 1950s, the computerization of machine tools took place. To make the task of setting and resetting the machine tool easier and quicker, the automatic mechanical controls were replaced with electronic logic circuits. Beside the machine tool stood a plug board (like a telephonist's switchboard) on to which the operator plugged in the sequence of operations required. This was the first version; it was called numeric control, or NC machines.

In the early 1960s, machine tools were joined to small computers. The sequence of instructions or program that the machine followed was stored on a paper tape. To change the sequence of operations, the operator had to change tape which was prepared by a programmer.

During the 1970s, the next step was to give the machine tool a built-in programmable computer with a conversational video display unit and keyboard. This removed the need for paper tape. The availability of small, reliable computer components in the form of the microprocessor expedited the development of what is now called computer numeric control (CNC) machines. These machines are equipped with adaptive controlling devices which can adjust their speeds and other operating characteristic by taking into account any adverse working conditions as they may arise during the manufacturing process, such as tool wear, torque deflection, etc. Human intervention at this stage is reduced to loading and removing work pieces and replacing dulled tools. If each machine tool can be operated and controlled by its own computer, it is possible to schedule and control the work of several machines from one central computer, which tells the operator, through a video screen, which work pieces to fit on to which machines. This technology was commercialised in the 1980s and was termed as direct numeric control (DNC).

Flexible manufacturing system: Flexible manufacturing systems (FMS) represent a new application for machine tools in which groups of machine tools are integrated and controlled by a central computer, the same computer also controls integrated materials handling systems, including robots, that move work pieces from machine to machine and position a work piece at each machine. A greater degree of automation than individual flexible manufacturing system could further become commercially attractive, particularly when there is a pressure from international competition to reduce the production costs. Turnkey automated factories have already been designed for industries, such as chemicals, cigarettes, paper, and textiles.

The prospects for automated factories in the metalworking industry are quite high, particularly where plants are to be based on the flexible manufacturing systems, and inventory management, scheduling and routing are to be controlled by computers. For example, at Cummins Engine Company, a manufacturer of diesel engines based in Columbus, Indiana, using the flexible system cut the time required to make the engines available in the market by half, reduced annual warranty expenses by an estimated $300,000, and reduced costs to the customers by more than 30 per cent. Further, robots and other automated equipment for such non-metalworking tasks as painting, controlled by the same central computer that coordinates metalworking production, are also becoming a reality. In addition, for both FMS and automated factories, the specifications of the parts to be produced can be developed by computer-aided design (CAD) equipment that will be connected to the rest of the production system. CAD then becomes the input of CAM and output of the CAM is the finished ready-to-sell product. Increasingly, therefore, new products will arise from the integration of information, electronic, and mechanical technologies. Although the design and installation of FMSs are high, they are more productive and efficient than a traditional manufacturing system.

Initialing control source	Type of machine response		Power source	LEVEL OF MECHANIZATION	
From a variable in the environment	Responds with action	Modifies own action over a range of variation	Mechanical-Nonmechanical	17	Anticipates action required and adjusts to provide it
				16	Correct performance while operating
				15	Correct performance after operating
		Selects from a limited range of possible fixed actions		14	Identifies and selects appropriate set of actions
				13	Segregates rejects according to measurement
				12	Changes speed, position, direction according to measurement
	Responds with signal			11	Records performance
				10	Signals pre-selected values of measurement. Includes error detection
				9	Measures characteristics of work
From a control mechanism that directs a predetermined pattern of action	Fixed within the machine			8	Actuated by introduction of work piece or material
				7	Power tool system, remote controlled
				6	Power tool, program control
				5	Power tool, fixed cycle (single function)
From man	Variable			4	Power tool, Hand tool
				3	Power hand tool
			Manual	2	Hand tool
				1	Hand

Figure 6.1 Level of mechanization

Source: Bright (1958)

In most countries, the market for complete flexible manufacturing systems still remains small relative to the economy as a whole. The importance of the FMS concept, however, goes beyond the number and growth of complete systems. More modest, partial, or limited applications of entire automated systems are widespread, consisting of production subsystems and incremental stages of planned FMS projects. However, from the seller's standpoint, the market for FMS components is considerably larger than the number of companies capable of purchasing a fully automated factory. Figure 6.1 presents the seventeen levels of mechanisation as identified by Bright (1958)

The Machine Tool Industry

Machine tool manufacturers are located throughout the UK but their concentration is more in West Yorkshire and both the West and East midlands with the average employment figure in 2002 being 8,600 compared with 10,200 in 2001 and 11,100 in 2000. Average percentage change in employment in the machine tool industry is more than average percentage change in durable goods industry. This can be attributed to the fact that the level of fluctuations in the employment conditions is sharper due to the cyclic nature of the machine tool industry. Further, the fragmented nature of the industry with relatively low level of capital investment and conservative management style makes it difficult to attract, train and retain bright engineering, managerial and technical talent. Table 6.1 shows the capital investment in the recent years.

Table 6.1 Private business investments in capital equipment 1991-2002

Year	Value
1991	13.33
1992	12.25
1993	11.45
1994	12.16
1995	14.32
1996	14.66
1997	16.20
1998	17.10
1999	14.64
2000	14.83
2001	14.54
2002	12.52

Source: Office of National Statistics, First Release (Series INHM)

The 2003 world machine tool survey suggests that most counties' exports fell, particularly Japan dropped more sharply than Germany. This worldwide reduction in export reflected in the reduction of British machine tool exports. Table 6.2 reports the export figure for the last decade. The figures for the UK imports and exports by products are presented in Table 6.3.

Table 6.2 Trade at current prices 1992-2002

Year	Sales of UK Goods	Exports	As a % of production	Imports	As a % of consumption	Implied consumption	Crude trade balance
1992	544	329	61	411	66	626	-82
1993	463	339	73	330	73	454	+9
1994	534	361	68	354	67	527	+7
1995	642	441	69	563	74	763	-122
1996	733	546	74	642	77	830	-96
1997	808	459	57	707	67	1057	-248
1998	696	538	77	698	82	856	-160
1999	542	510	94	520	95	552	-10
2000	536	464	87	574	89	647	-111
2001	573	492	86	543	87	623	-50
2002	392e	369	94e	437	95e	459e	-68

Source: Office for National Statistics, HM Customs & Excise and MTA calculations, e = estimates

Table 6.3 UK imports(£436.8m) and exports (£369.2m) by product type, 2002

Product	% of import	% of export
Physico-chemical	12.2	13.6
Machining Centres	11.7	20.6
Lathes	15.4	12.7
Milling, Drilling and Boring	12.5	6.5
Grinding and Finishing	11.1	15.5
Special Purpose Machines	2.6	3.8
Other metal cutting	6.7	6.1
Bending	8.4	6.3
Presses	12.9	7.7
Other Metal Forming	6.4	7.2

Source: HM Customs & Excise, Business & Trade statistics and www.uktradeinfo.co.uk; calculated by MTA

Tables 6.4 and 6.5 report the data for leading export and import markets for the year 2000-2002, respectively. The 2003 machine tool survey reports that China is the biggest market, out-buying Germany by a difference of US$ 880 million (equivalent to what Canada buys). Further, China's appetite as the biggest consumer also makes her the number-one importer. Overall, Japan has the most favourable balance of trade in machine tools.

Table 6.4 Leading export markets 2000-2002, by country of destination

2000 Countries	value	2001 Countries	value	2002 Countries	value
All countries	463.9	All countries	492.2	All countries	369.2
European U	240.0	European U	243.3	European U	154.5
1 USA	87.8	1 USA	104.3	1 USA	55.8
2 Germany	64.6	2 Germany	65.3	2 Germany	43.3
3 Belgium	54.5	3 Belgium	56.7	3 Belgium	29.3
4 France	33.3	4 Spain	27.4	4 France	19.2
5 Ireland	21.1	5 France	21.7	5 China	17.3
6 Italy	15.8	6 Italy	20.5	6 Italy	16.0
7 Spain	13.7	7 Ireland	15.2	7 Japan	13.5
8 Japan	12.9	8 Sweden	12.7	8 Poland	13.2
9 Sweden	11.7	9 Czech	12.6	9 Ireland	12.4
10 Canada	11.1	10 China	12.3	10 Spain	10.3

Source: HM Customs & Excise via www.uktradeinfo.co.uk
Figures include export of new and used machines. Values are in £ millions.

Table 6.5 Leading import markets 2000-2002, by country of origin

2000 Countries	value	2001 Countries	value	2002 Countries	value
All countries	574.5	All countries	542.6	All countries	436.8
European U	242.2	European U	206.7	European U	190.8
1 Germany	144.6	1 Japan	126.0	1 Germany	112.6
2 Japan	119.0	2 Germany	112.8	2 Japan	107.8
3 USA	84.7	3 USA	80.7	3 USA	44.9
4 Switzerland	44.7	4 Switzerland	38.1	4 Italy	24.5
5 Taiwan	31.1	5 Taiwan	33.0	5 Switzerland	23.3
6 Italy	23.2	6 Belgium	31.3	6 Belgium	20.4
7 Belgium	23.0	7 Italy	17.6	7 Taiwan	14.4
8 France	16.3	8 S. Korea	16.6	8 Canada	9.9
9 S. Korea	13.4	9 France	11.8	9 S. Korea	9.8
10 Spain	10.7	10 Spain	10.1	10 France	8.6

Source: HM Customs & Excise via www.uktradeinfo.co.uk

Figures include import of new and used machines. Values are in £ millions. Import figures from countries of the European Union are based on country of consignment.

According to the figures complied by Gardner Publications, the top 29 countries produced about US$ 31 billion in 2002 which is below 14 per cent of the value those same nations created in 2001. Japanese machine tool production reduced by one-third, placing it behind Germany. The USA also dropped by one-third, making a pave for China to move ahead. The UK machine tool industry was estimated to be 11th in the world league table for production. The market shares of nations ahead of the UK are presented in Table 6.6.

Table 6.6 Share of world production of machine tools (US$ 31.0bn)

Country	% share	Country	% share
1 Germany	21.7	7 Switzerland	5.5
2 Japan	20.6	8 Spain	2.8
3 Italy	12.2	9 S. Korea	2.7
4 China	9.8	10 France	2.6
5 USA	6.2	11 UK	2.0
6 Taiwan	5.7	12 Others	8.4

Source: Gardner Publications Inc.

Table 6.7 reports the figure for the world exports of machine tools. The UK remained the 7th largest exporter in the world in 2002, Germany and Japan being the largest and the second largest exporter of machine tools.

Table 6.7 Share of world exports of machine tools (US$ 17.3bn)

Country	% share	Country	% share
1 Germany	22.4	7 UK	3.1
2 Japan	19.8	8 Spain	2.9
3 Italy	10.4	9 France	2.5
4 Switzerland	8.5	10 Belgium	2.4
5 Taiwan	8.3	11 S. Korea	2.3
6 USA	5.3	12 Others	12

Source: Gardner Publications Inc.

Conclusion

In the past, growth in the machine tools industry has been primarily in the automotive and aviation sector. This is no longer the case. Now the growth in machine tools is a function of the worldwide industrial climate, particularly because international competition is causing manufacturers worldwide to reduce overhead costs and improve the quality of products. To accomplish this, most manufacturers are modernising manufacturing units by installing computer numeric control (CNC) machines, flexible manufacturing systems (FMS) and flexible manufacturing cells (FMC). The result is that production costs are reduced and quality is improved, both of which are critical to survival in manufacturing industries.

Plant modernisation is taking place in all industries. Automotive and aerospace sectors are the pace setters, but appliance, electrical and electronics industries are right on their heels. High standards of performance for appliances, often established by regulation, are forcing the appliance industry to invest in high-technology, such as CNCs, FMSs, etc. Increased sales and high turnover rate in the electrical and electronics industries are forcing manufacturers to increase their production capacity.

In the future it appears that CNC machines will have more advanced features through the introduction of new CNC controllers that can be: integrated into a system rather than operate standalone, i.e. networking of CNC machine tools; developed based on ease of programming, graphical user interface (GUI) environments and blueprint programming (following a set of simple prompts that will generate a blueprint emulation on the screen); and employed in a parallel processing architecture to reduce block processing time, speeding up set up and operational functionality.

Indeed, the machine tool industry has undergone fundamental changes over the past decade. Although basic metal cutting and metal forming machines are still a critical element in the manufacturing industry, the machine tool industry today has become part of a new, automated manufacturing industry that produces new types of products, such as computer-driven, integrated production systems, that perhaps rarely existed a decade ago. Keeping in touch with the evolving technology, it is this automated market that needs to be tapped by the advanced machine tool manufactures. It is this new advanced technology-based market that forms the basis for an assessment of the machine tool industry's responsiveness to domestic and foreign needs in sectors such as defence, aerospace, automotive industry, and others.

Chapter 7

Case Studies

Introduction

The purpose of chapter is to explore the market orientation practices in the select machine tool companies, to understand the process of becoming a market-oriented company, and to examine the level of market orientation perceived by suppliers and their respective customers.

The focus of each case study is at corporate level. Each case represented assimilation of managerial thinking under a unique set of business conditions, such as market turbulence, technology turbulence, competitive intensity, competitors' concentration, among others, and therefore a unique emerging pattern of market orientation for each company was observed. The emphasis of the cases is on the understanding the corporate executives' attitude towards their customers rather than on any functional department of the business. The case study is restricted in that sense. However, the marketing stance was further explored by conducting a limited survey. A gap analysis was performed to detect differences between perceived attitude of managers and their respective customers towards market orientation.

Qualitative Research Design

Qualitative research meets different objectives from quantitative research, and provides a distinctive kind of information. For industrial research purposes it may be carried out with some kind of linkage to statistical enquiry (i.e. to help develop, illuminate, explain or qualify statistical analysis findings), or it may be entirely independent. Either way, it is important that the contributions of qualitative research are fully exploited. If decisions or actions are to be based on qualitative research, the managers need to know how the findings of the research have been obtained. This process would not only foster greater confidence in the findings, but also a deeper understanding of what qualitative research can accomplish.

Further, qualitative data analysis is about detecting, categorising, theorising, exploring and mapping of the critical issues being analyzed. Therefore, the selection of the research method should be such that it allows certain tasks to be performed, although these tasks may depend upon the research questions being addressed. Case studies can be used to accomplish various tasks such as to provide description (Kidder, 1982), to test theory (Anderson, 1983; Pinfield, 1986) or to generate theory (Harris and Sutton, 1986; Gersick, 1988). This chapter concerns providing descriptions and the testing of theory. Attempts were made to understand the top executive's attitudes towards market orientation theory, to find association between quantitative and qualitative findings about customers' perception of their suppliers' attitude towards market orientation, and to seek explanation for the findings. A review of literature discusses several aspects of case studies. For example, Glaser and Strauss (1967) detailed a comparative method for developing grounded theory, Miles and Huberman (1984) codified a series of procedures for analyzing qualitative data, and Yin (1981b, 1984) described the design of case study research. In keeping with the objective of the chapter, this study is designed to facilitate the systematic analysis within the framework suggested by Yin (1981a, 1981b). The case analysis framework is presented in Figure 7.1.

For the purpose of the case analysis, the definition of case study given by Yin (1984) has been adopted. It states that a case study is an empirical enquiry that investigates a contemporary phenomenon within its real-life context when the boundaries between phenomenon and context are not clearly evident and in which multiple sources of evidence are used. Further, Yin (1984) suggests considering the following components of a research study: a study's question and proposition, its unit of analysis, the logical linking of the data to the propositions, and the criteria for interpreting the findings. These components are discussed in the next section.

The Study's Question and its Proposition

The premise of this study is that market orientation is positively related to business performance. Second objective is to examine if the level of market orientation reported by suppliers is the same as their customers. If an organization is market oriented, then the evaluation of how market oriented that organization is should also come from its customers. This is a critical point. In other words, if an organization is successful in meeting the needs of its customers, the gap between the level of market orientation reported by both suppliers and buyers should ideally be zero. Bolton and Drew (1991) emphasized the need for looking through the eyes of customers because they are likely to define problems, and hence solutions. While it is recognized that suppliers' and customers' perceptions about market orientation might not be the same, the following proposition is put forward - The customer's perception of the market orientation is more important than suppliers' own perception of market orientation in explaining the suppliers' own business performance.

Unit of Analysis

The study takes individual organization as the unit of analysis for the case study with an objective to explore the actions taken by individual firms towards market orientation. Further decisions were made about the collection of data and the specific time boundaries needed to define the beginning and the end of the case study. All of these questions were considered and answered to define the unit of analysis.

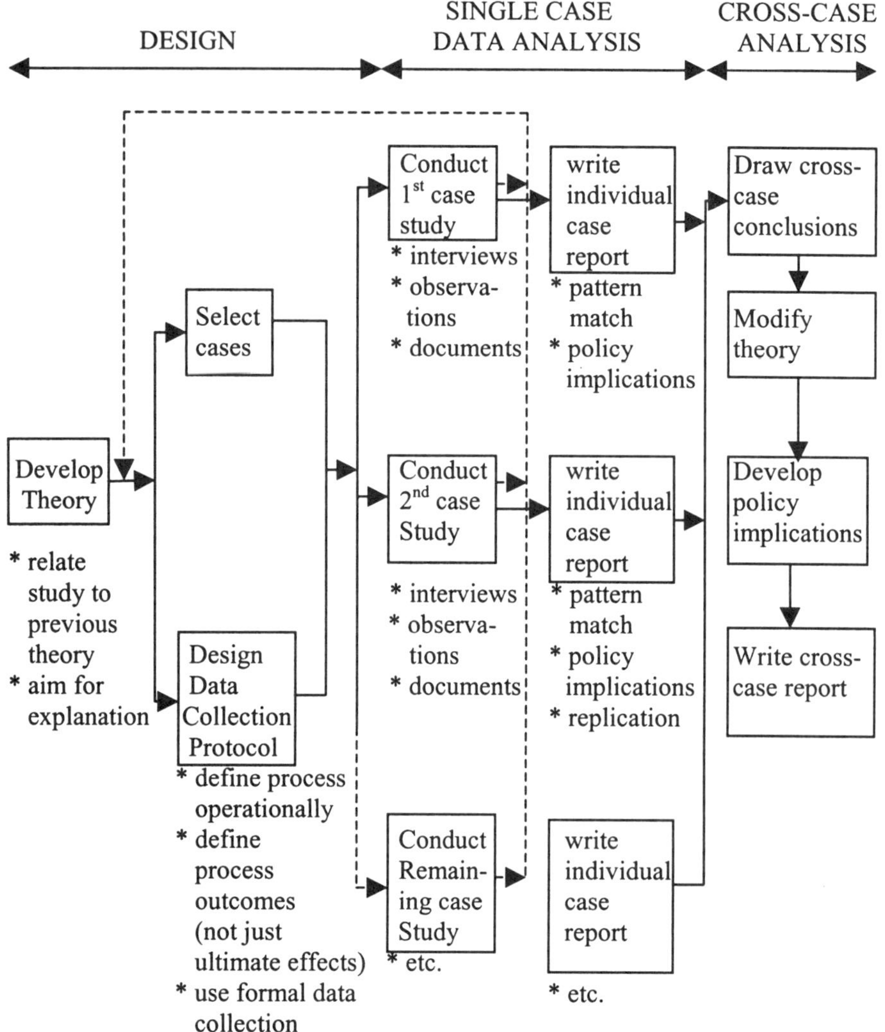

Figure 7.1 The framework for case analysis

Source: Adapted from Yin (1984)

Selection of companies: The case study followed the recommendations and framework of Greenley (1988) and Yin (1984) respectively. The selection of the cases was based on the following criteria: the companies were registered either under SIC 3541 or SIC 3542 to ensure that companies belonged to the machine tool industry; companies had a marketing department as a separate function of the business with delegated responsibilities for marketing, whose managers practised the concept of market orientation and were aware of the outcome of such practices.

During the mail survey, it was learnt that there were 12 companies that satisfied the above criteria. Consistent with Pettigrew's (1988) deliberate theoretical sampling plan methodology, selection of these companies was based on two parameters: level of their market orientation and sales turnover. To select the companies objectively, turnover and market orientation variables were split by median into two groups. For example, companies having score for market orientation below and above median were considered as low and high market-oriented companies, respectively. The medians for turnover and market orientation are 5.2 and 5.5, respectively. Turnover is the annual sales of the company in millions whereas market orientation is a score on a seven-point scale. Of the 12 companies only four companies were selected for the development of the case study. The notion of selecting a few companies is supported by Pettigrew (1988) who noted, given the limited number of cases which can usually be studied, it makes sense to choose cases, such as extreme situations and polar types in which the process of interest is 'transparently observable'. Mintzberg (1979, p. 585) echoed the same sentiment, 'no matter how small our sample or what our interest, we have always tried to go into organizations with a well-defined focus – to collect specific kinds of data systematically'. The select four companies are reported in Table 7.1.

Table 7.1 Matrix for the selection of companies

	HI Market oriented firms (> 5.5)	LO Market oriented firms (< 5.5)
LO turnover firms (< £5.2 m)	Company A (TO = £5 m, MO = 5.81)	Company C (TO = £3 m, MO = 3.65)
HI turnover firms (> £5.2 m)	Company B (TO = £60 m, MO = 6.2)	Company D (TO = £6.5 m, MO = 5.24)

TO=Annual Sales Turnover, MO=Market Orientation on a seven-point scale

While selecting companies, care was taken to ensure that each company represented some key manufacturing differences. For example, Company A specialized in metal cutting type machines, such as band saws, circular saws and power saws; Company B was one of the oldest companies engaged in the production of high quality machine tools and computer numeric control machines; Company C was one of the biggest and largest supplier of lathe machines; and Company D concentrated on the production of innovative cutting tools. Attempts were made to represent the variety of machine tool manufacturers in these four cases. Further, the selection of these four different companies from the same industry allowed the researcher to control for environmental variations, while focusing on the variations of market orientation of companies due to characteristics, such as size of the companies (small vs. large), nature of the products (basic vs. high-tech tools), form of companies (manufacturer vs. distributor), and others. For confidential and commercial reasons, the names of the companies have been disguised and their balance sheets have not been disclosed.

Linking Data to the Proposition

This objective was achieved by developing a semi-structured interview questionnaire. Managers were asked questions relating to market orientation, such as what does the term market orientation mean to you? What kind of things does your company do to be market oriented? What are the consequences of being market oriented? The objective was to explore how each company had been trying to be market oriented. Managers were asked about a situation in which the market orientation concept was not appropriate. Further they were asked to discuss their business performance as a result of market orientation. The formal interviews lasted between half an hour and an hour. In one case it lasted up to three hours including a trip to the manufacturing unit.

In addition to the semi-structured interview, development of a questionnaire was thought to be a suitable technique for the following reasons: First, responses obtained through questionnaires maintained some degree of uniformity, which reduced the chances of unreliability caused by an un-structured interview technique; second, the uniformity in the structure facilitated detection of patterns in the responses; and third, the concise questionnaire helped gain insight into any parallel conclusions that occurred across certain situations.

The Likert-type questionnaire was used to match pattern between suppliers and their customers on key variables as suggested by Campbell (1975). Because the potential patterns should be compared against each other and should be related to the available data to examine which one of the patterns best describes the data, the customers of each company were asked to respond to the same questions as did their respective suppliers. The Questionnaires were slightly adjusted and rephrased to make it suitable for customers to respond (Harris and Sutton, 1986). Such adjustments allowed the researcher to probe emergent themes relating to market orientation and to take advantage of special opportunities which might be present in a given situation. Additional financial information was also collected where applicable.

Managers chose randomly the names of their customers. Of these customers, I contacted four customers from each company. They were assured of their anonymity. Each customer of these companies received a letter outlining the purpose of the study along with a questionnaire and a prepaid envelope. Three weeks later another set of questionnaires was sent to those who did not respond. Of 16 customers, 11 sent us their complete questionnaire. Following the collection of data, a gap analysis was performed to compare the scores between the responses of suppliers and their customers on the statements presented in Table 7.2.

The Criteria for Interpreting the Findings

A number of researchers have undertaken their own variations and additions to the

Table 7.2 The matched questionnaire for customers

Overall how true are the following statements reflective of Company "X" as a supplier.

1 Company X reveals our needs through customer research. (M)
2 Company X produces and offers products that are truly satisfying our needs. (M)
3 Company X commits resources in research and development to ensure the quality of the products suits our needs. (M)
4 In practice, Company X avoids our conflicts between market orientation and profit orientation. (BD, rc)
5 We are aware of how the entire business of company X can contribute towards creating a value for us.
6 Company X keeps its promises after sales and delivery. (R)
7 Company X is prepared to disagree with us in order to achieve a better decision for us. (SW)
8 Company X tries to influence by information rather than by pressure. (SW)
9 Company X's representative normally meets us once a year to asses our needs. (JK, rw)
10 Representatives from Company X's manufacturing department interact directly with us to assess how best to serve us. (JK)
11 Company X frequently polls our end-users to assure the quality of its products and services. (JK)
12 Company X is slow to detect fundamental shifts in the British machine tool industry, e.g. competition, technology regulations. (JK, rc)
13 Company X tends to ignore changes in the product or services our organization needs. (JK, rc)
14 Company X reviews its product development efforts to ensure that they are in line with our organizational needs. (JK)
15 Company X normally takes corrective action immediately for any complaints. (JK, rw)
16 Company X is market oriented. (the author)

M=Meziou (1991), BD=Barksdale and Darden (1971), R=Ruekert (1992), SW=Saxe and Weitz (1982), JK=Jaworski and Kohli (1993), rc=reverse coded, rw=re-worded.

earlier methodological work (Harris and Sutton, 1986; Gersick, 1988; Leonard-Barton, 1988). Some of them have employed their own techniques for building theory from cases (Jick, 1979; Sutton and Callaham, 1987; Bourgeois and Eisenhardt, 1988; Van Maanen, 1988). In order to achieve the objective of the study, Campbell's (1975) technique that the data should match the pattern was employed; however confusion surrounded as to how close the match should be in order to be considered a match. Consistent with the pattern matching methodology, no statistical tests were performed. For pattern matching purposes, randomly selected customers from each company were contacted. Although there is no precise way of setting the criteria for interpreting these types of findings, if the different patterns are sufficiently contrasting, then the findings can be interpreted in terms of comparing at least two rival propositions (Yin, 1984). Several researchers have relied on tables which summarize and tabulate the evidence underlying the statements of a construct (Miles and Huberman, 1984; Sutton and Callaham, 1987). I created a table for each case study to explore the proposition that perceptions of suppliers and their customers about market orientation were different. A subjective (Campbell, 1975) difference of more than ± 1.0 between the two means was used for a statement to be significantly different from each other.

Further, to judge the quality of any research design, Kidder (1981) recommends four types of tests: construct validity, internal validity, external validity and reliability. As a research design is supposed to represent a logical set of statements, the quality of the design can be tested by certain logical tests. Because of the small sample size (I had four suppliers and 11 customers), statistical tests could not be conducted to demonstrate the reliability and validity of the findings obtained from the analysis of each case; however attempts were made to discover the pattern of agreement or disagreement between suppliers and their customers on statements presented in the questionnaire. If a pattern from one data source is corroborated by the evidence from another data source, the finding may be taken as reliable and better grounded. On the contrary, if evidence conflicts, the researcher may like to reconcile the evidence by probing further the source of differences.

Company A

This company dates back when five gunsmiths manufactured 200 snapshots muskets per month at 17 shillings per piece! At the time of the Crimean War, 14 gunsmiths decided to make their trading position formal, and in 1661 took the decision to form a company to manufacture guns. In 1921, the company set up its own machine tool division and later expanded into the production of bicycles, motorcars and motorcycles. The machine tool division produced single- and multi-spindle automatic lathes, copy and multi-tool lathes, grinding machines, transfer lines and broaching machineries. With the arrival of the electronic age, this company saw the need for augmenting the line of products. As a result, it launched a range of Batchmatic machines, which proved to be highly successful worldwide. Continuous development of this product range led to the addition of CNC machines, further enhancing the capabilities of these machines.

In 1992, the business expanded by acquiring other company's manufacturing rights to the 'Alpha' range of CNC turning centres and by translating the technology into the production of 'Beta' machines. A year later, the company acquired the manufacturing rights from another company to manufacture 'Gama' CNC multi-slide lathe. In 1994, the company further expanded by purchasing the assets of a company which held an excellent reputation for rebuilding the multi-spindle automatics.

The company had close associations with the British Standards Institute, the National Physical and Engineering Laboratories, the Machine Tool Technologies Association, world training schemes, and Universities. Major European research projects organised by the Advanced Manufacturing Technology Research Institute (AMTRI) had this company as a partner.

This company was recently given a facelift by modernising its plant and equipment with a complete refurbishment of the buildings. Now, it has extensive design, sales, purchase, and service areas to complement the manufacturing facilities.

Market Orientation of the Company

As the British machine tool industry was a mature industry, most of the companies were niche players. This company was one of the niche players, which thrived on producing low volumes of high quality machine tools. Products ranged from basic and standard to highly engineered and complex machines. Company A claimed to be market-oriented by paying particular attention to customer support, reliability of machines, availability of spare parts and wide application of software engineering. The company, by providing two reasons, supported the notion that there was a positive relationship between market-orientation and business performance. First developing nations were capable of reaching world quality standards; therefore, the company had to be more technically capable to produce machines to be more profitable than its counterparts who were less technically capable; and second, to be profitable, this company must be able to produce machines quickly to save on labour costs. Although the rate of technology change was high, the company was not adversely affected by this factor because the company entered into a contract when the design of a new machine was conceptualised and the latest technology was incorporated into the design throughout the manufacturing process. Therefore, the company ensured that it did not get caught in a 'technology vacuum'. The lead time for the company to manufacture machines was eight to ten months, which was quick enough to keep pace with the changing technology. This company boasted to have 50 per cent of its business from existing customers. Most loyal customers came from Australia and Mexico because of the ability of the company to backup application, customer support, and spare parts throughout the world through its agents. This company was a preferred global supplier to Ford. A detailed market oriented strategy of the company is offered in the following sections.

The Quality Control

The company placed the highest priority on the quality of the machine tools. All machines, parts and services supplied or manufactured by the company, including the bought-out proprietary items, made-out details, sub-contracted units and services, were covered by the company's Quality System and the requirements of International Standard BS EN ISO 9001. This company was approved by the

British Ministry of Defence and was endorsed by several major British defence manufacturers.

Product Range and the User Friendly Machines

The company had the widest range of turning machines manufactured in the UK. The Alpha offered a wide range of CNC turning centres, and the new generation Beta machines incorporated the modular design-and-build capabilities. For example, Beta's modularity included features, such as C-axis, driven tools, auto tailstock or even second spindles. The Beta machine with 42, 65, or 80 mm bar capacities in single- or twin-spindle configurations, offered an advanced range of module-designed and built CNC lathes. The company's range of multi-slide automatic lathes added a new dimension to the bar lathe technology. The 2600, 4200 and 6500 models, which suited all applications from 3 mm to 65 mm bar capacity, included features, such as eight-station turrets, and to minimize cycle times, they were equipped with control systems that calculated the optimum position for indexing, taking into account the length of both the withdrawing and advancing tools. Further, the turret was bi-directional with a 0.3 second indexing time between stations.

The PC control system was specifically developed for the company's multi-slide lathe. Any experienced setter could feel at home with its operation within half a day and no specialised programming knowledge was necessary. A program could be written to produce a component exactly the same way as a cam auto machine, but was completed on a screen in an average of 45 minutes rather than in six hours on a machine using cams.

The product range also included multi-spindle bar and chucking automatics, controlled by CNC or CAM systems, single-spindle bar automatics and multi-spindle bar automatics. The multi-spindle machines offered economic production repeatability, fast set-ups, reliability and simplicity whereas the single automatics evolved through many years of successful development of automatic machines to suit all applications from 3 mm to 50 mm bar capacity. By using an outside feeding attachment, the single automatics could be used for a bar up to 60 mm in diameter. Other six-spindle bar automatic machine had a simplified drive with infinitely variable speeds and feeds, higher spindle speeds and facility for tool life monitoring to all stations. The simplified stroke setting made these machines user friendly.

Export Orientation of the Company

The company paid close attention to the growing export markets and offered a range of machines to foreign markets. These machines were now being shipped around the world to manufacturers, making the company thrive in a wide variety of industries. Current overseas orders included Mexico, Spain, Germany, Australia, Sweden, Italy and China. The company had offices in India in Banglore, Bombay, Pune and Madras, which were managed by a local agent. Further, the company had joined the MTTA initiative in India, with an office in Pune, which represented the company and other British manufacturers. The Indian market was projected to become one of the largest in the world for the import of quality machine tools, so the company was well equipped to export the high quality machines. Additional, export orders included an automated cell for a manufacturing plant in Sweden, comprising the Alpha, the Beta, 8-axis twin-spindle turning centre fitted with a gantry and a post-process gauge. In China, a local Diesel plant had taken a delivery of the Alpha's four Series CNC lathes fitted with the pro 90 out of round turning feature. The Diesel Company, based in the famous city of the ancient Terra Cotta Army, was the manufacturer of marine engine pistons with varying oval along the longitudinal stroke. The international business continued to grow with orders placed from Mexico and Australia for the company's multi-spindle lathes. Perhaps the simplest mode of internationalising a domestic business is exporting, particularly because little or no investment is needed to have an exposure to the host country.

Customers' Perception of the Market Orientation

The completed questionnaires were used to perform a gap analysis for between the company and its customers on the perceptions of market orientation. The scores are summarised in Table 7.3.

Table 7.3 Means for the company A and its customers

Items	Company (A)	Customers (B)	Differences (A-B)
1	5.00	3.33	+1.67*
2	5.00	6.00	-1.00
3	6.00	5.67	+0.33
4	4.00	4.00	0.00
5	3.00	1.33	+1.67*
6	6.00	3.33	+2.67*
7	7.00	4.33	+2.67*
8	6.00	5.33	+0.67
9	7.00	4.00	+3.00*
10	5.00	3.33	+1.67*
11	5.00	2.33	+2.67*
12	7.00	5.33	+1.67*
13	2.00	5.33	-3.33*
14	6.00	3.33	+2.67*
15	6.00	5.33	+0.67
16	4.00	5.03	-1.00

* = a difference of more than ± 1 was assumed as being significant.

Conclusions and Discussions

Table 7.3 indicates that the differences in the means between the company and its customers are not significant for the items: 2 (-1, products satisfy customers' needs), 3 (+0.33, supplier ensures the quality of the products to suit customers' needs), 4 (0, there is no conflict between market orientation and profit orientation), 8 (+0.67, the supplier tries to influence its customer by information rather than by pressure), and 15 (+0.67, the supplier takes corrective action immediately for any complaints). However there are significant positive differences on items 1, 5, 6, 7, 9, 10, 11, 12 and 14, suggesting in general that the company's perception of market orientation is higher than those perceived by its customers.

Although, there are differences between both perceptions, the difference on the statement 16 suggests that the overall market orientation of the company perceived by its customers is almost same as the perception of the company. The major difference was observed on item 9 (the difference being +3) indicating that the company's representatives do not meet its customer once a year to assess their needs. This difference may be due to the fact that the select customers in the sample were those who were not contacted once a year. Another reason could be that the particular respondents of the questionnaire were not aware of the visits of the company's representatives. This may be particularly true if the company is large.

On the contrary, statement 13 (the difference being -3.33) that the company does not ignore changes in the product or services its customer needs is noteworthy. It is indeed a good sign for the company to be market oriented. In fact, customers were satisfied with the changes made by the company to its products to meet customer needs, which was further confirmed by statement 2 (the difference being -1) that the company produces and offers products that truly satisfy its customers' needs. The company is well established to meet the challenges of the future.

This company has grown exponentially by the introduction of new products and acquisition of other companies' product range to meet its customer's needs. This had led to a considerable expansion in business compared with only a few years ago when this company was primarily engaged in the design, manufacturing and servicing of machine tools. Now products vary from the well established single- and multi-spindle automatics to sophisticated CNC lathes with twin-spindle and power tooling. These features proved to be attractive to not only overseas markets, but also to the aerospace and automotive industries. Thus, the company offers a range of machines to suit a range of machining operations, which incorporates advanced technology that are well suited to take quality manufacturing into the next millennium.

Company B

In the early 1990s, the British machine tool industry was hit by a recession that saw the world market down by nearly 50 per cent. At that time this company was one of the two companies under a group of companies, responsible for the manufacturing of lathe machines. Both these companies were running at barely half of their capacities. The difficulty faced by the Director was that each company was quite different from the other. The problem of one company was that it was technology-led, whereas the other company was sales-led. As a result, both companies failed to satisfy customers' needs. The board of directors of the group was adamant that both companies should survive, which meant that it was faced with making some harsh decisions. As there was no future in having two companies running at half capacity, it was thought that it was prudent to have one company running at the maximum capacity. Directors also felt that the recovery could only be achieved through a market-led strategy. Following this decision, these two companies were combined together to form a new company 'B', and a new marketing director with a proven track record in marketing and product development was hired to implement the market-led strategy, i.e. the first step towards market orientation.

Market Orientation of the Company

The Marketing Director defined market orientation as market led skills necessary to meet customer needs. This company was one the largest lathe machine manufacturers in the UK, which had customers from UK, USA, Europe, and rest of the word in the proportion of 20, 30, 30 and 10 per cent, respectively. The grand success of the company's newly designed machine 'Theta,' which was based on market research, confirmed that the company was market oriented. This company took market research activities seriously; all information pertaining to customers were communicated across all departments. During the interview with the Director of the company, it was learned that in order for the information to flow freely from the marketing department to the shop floor, the marketing department was located inside the manufacturing plant. To make the information flow smoothly across departments, the commercial manager sat with the sales department in the same building, which was an example of being market oriented in terms of the free flow of information pertaining to customers across other departments. This company practised the concept of market orientation and had been successful, as the sales of the machines exceeded expectations. Company B believed in the positive relationship between market orientation and business performance. On asking the Managing Director, if they could envisage a situation in which the relationship may not be profitable; for example, high rate of technology change, market turbulence, or competition could have prevented his company from being market oriented. The reply was that the relationship between market orientation and business performance was robust, and should not be affected by environmental factors like competition, market turbulence or technology turbulence. This implied that in order for a company to be profitable, it must meet the needs of its customers irrespective of the business environment. However it was acknowledged that it might not be feasible to have expertise enough to be fully market oriented, but one can try to satisfy the need of customers given the amount of resources and expertise within a company. This company conducted market research on a regular basis to identify if the key attributes of a machine were being utilized effectively. If not so, the reasons were discovered and corrective actions were put into place immediately. The unwanted attributes were removed in the development of the next batch of machines and prices were adjusted accordingly. In this respect, market orientation led to lower costs and simpler machines with fewer frills. The in-house market research team of the company conducted about ninety per cent of the market research by interviewing end-users of its machines. As a result of the implementation of the marketing concept, the company grew its global market share from six percent to 36 per cent in the last five years. A detailed market orientation program of the company is discussed in the following sections.

The Marketing Research

The first step taken by this director was to conduct massive market research to reveal the needs of the customers so that a new machine could be developed to match the expectations of their customers. Following the interview with the Director, it was discovered that in the UK more than 65 per cent of all engineering shops had a lathe machine manufactured by the company; therefore, these exiting customers should contribute first to the detailed market research exercise. More than 3,500 questionnaires were mailed to establish factual details relating to the technical specification of machines, such as size, accuracy and the number of hours worked per week on a particular machine. This was followed by 300 telephone interviews worldwide to establish potential user preferences relating to details, such as colour, country of manufacturer, after-sales requirements, preferred methods of programming and preferences for sources of CNC controllers. While face-to-face interviews confirmed that the company's brand name was not well associated with CNC lathes, it did confirm that the market was willing to try a new kind of machine. Awareness testing results showed that the company was regarded as the number one lathe supplier in the UK and abroad. Further, this testing also established that the company had a good reputation for good quality and value-for-money, and that this company's lathes were also considered to offer good resale value. However the market research also made it clear that many customers would not accept this company's new machines as a me-too supplier. The research also revealed that this company's products were wrongly positioned. So the company's prime job was to either reposition the existing product or develop a new product, and then position it where it could target a larger market to achieve growth. This fact influenced the company's move into the production of a low cost, no-frill, two-axis lathe. As a result, the company established an acceptable selling price for the machine (Theta) rather than using this company's name as the prime reference.

The Product Specification and the Pricing

The analysis of the in-house data relating to the customers revealed that three different sizes of machines were needed. The data also established precise machine specifications and exact corresponding sales prices. Knowing the detailed specifications, now it was the job of the company to introduce the product to the market at a competitive price. Although it was not an easy task, the company was adamant that there was absolutely no scope to deviate from either the technical specification or the costs now established, because research confirmed that customers were invariably recession-damaged and only prepared to pay the prices determined by the market research. The emphasis on price also meant that this company had to be able to compete with good quality used machines because nobody wanted to buy a used machine when a customer could have a new one for the same price.

As a result, the engineering team met on a weekly basis to discuss these ideas. Each designer was given a strict cost and performance target to reach. If this target was not reached, they had to redesign their area of responsibility until it was acceptable, not only in terms of cost but also the engineering design, stress analysis (FEA, Finite Element Analysis) and computer modelling tests. For major parts, such as turrets, CNC controls and guarding, the company established a strategic alliance and a long-term commitment that resulted in prices 50 to 60 percent cheaper than previously bought. To control the cost, the company also decided that commodity items (standard specification items) could be bought from any source worldwide which met the cost constraints, technical performance and delivery requirements.

By conducting the in-house market research, three models of machines were developed. By redesigning the components of the machines, the costs of the production for these machines were significantly reduced. Machine guarding was another innovation to bring the cost down, as this was designed, in conjunction with another guarding manufacturing company, to be dropped straight onto the machine, requiring no further work. Another major innovation was the concept of moving a conveyor line for machine assembly. The new manufacturing system did not only force a near-perfect flow of the work but also a quality discipline because the line could not be stopped; parts had to be there on time, and had to be right.

Customers' Perceptions of the Market Orientation

The completed questionnaires were examined to test for the differences between the perceptions of the company and its customers about the market orientation. On the next page, Table 7.4 reports the means for both groups.

Conclusions and Discussions

Table 7.4 suggests that the positive significant differences between the perceptions of the market orientation between the company and its customers are on the items: 1 (+2.67, the supplier reveals customers needs through customer research), 3 (+1.67 the supplier commits resources to ensure the quality of the product suits its customers' needs), 6 (+2, the supplier keeps promises after sales and delivery), 11

(+4, the supplier frequently polls its end-users to assure the quality of its products and services), 13 (+2.67, the supplier ignores changes in the products or services that its customer needs), 14 (+ 3, the supplier reviews its product development efforts to ensure that they are in line with their customers' needs), 15 (+ 2, the supplier takes corrective action immediately for any complaints), and 16 (+2, the supplier is market oriented).

Table 7.4 Means for the company B and its customers

Items	Company A	Customers B	Differences A-B
1	7.00	4.33	+2.67*
2	7.00	6.33	+0.67
3	7.00	5.33	+1.67*
4	4.00	3.67	+0.33
5	4.00	4.33	-0.33
6	5.00	3.00	+2.00*
7	4.00	4.67	-0.67
8	5.00	5.67	-0.67
9	7.00	6.00	+1.00
10	7.00	6.00	+1.00
11	7.00	3.00	+4.00*
12	1.00	5.33	-4.33*
13	7.00	4.33	+2.67*
14	7.00	4.00	+3.00*
15	5.00	3.00	+2.00*
16	5.00	3.00	+2.00*

The significant positive differences between the perceptions were detected in the statements: 11 and 14. These two items are concerned with the quality of the company's products and services and with the alignment of products with their customers' needs. The differences were attributed to the fact that the company had a reputation for manufacturing lathe machines and not CNC machines. Therefore, customers were unsure about the quality of the new machine, as it was too early to know the performance of these machines. The director of the company admitted that it was not possible to satisfy all of its customers' needs. However the company was successful in satisfying the need of its 90 per cent customers.

It is interesting to note the negative difference in the mean on item 12 (the difference being -4.33), suggesting that the company was slow to detect fundamental shifts in the machine tool industry, e.g. competition, technology regulations, whereas its customers perceive otherwise. No wonder the company attempts to be market oriented under all conditions of business environment.

Although it is not conclusive that the company is being perceived as market oriented by its customers, the success of the newly designed machine was confirmed by the sales which exceeded the expectations of the company.

Following the launch of the new machine in April 1994, the company sold 800 machines in the first year. In a little over two years, it had sold more than 1,800 units, while for the financial year from April 1996 to 1997 it manufactured more than 1,500 machines, including 500 machines for export to the US. In fact, 80 per cent of the machines were being exported through a worldwide network of 55 distributors.

Company C

This company's reputation was founded on supplying an unrivalled range of quality machines and systems suitable to meet every conceivable sawing and storage need. The company had installed thousands of sawing systems, which were supported by the company's proven ability to maintain equipments after sales. In its 30 years of history, the company had supplied machines to virtually every industrial sector by expanding the range of select products; for example, from major retailers, like Tesco, to manufacturers, such as British Aerospace and JCB Excavators.

For decades, the company's principal had set the standards for the manufacturing of high quality machines and pioneering a host of new sawing and storage techniques. For example, the innovative application of microelectronics and the development of polymer concrete construction kept the company ahead of the competition. The company had worked for 20 years with the principal, which specialized in the production of sawing techniques, such as power hacksaws, band saws, circular sawing machines, fully automatic sawing machines and storage systems for engineering applications. In the UK, a total of 24 such systems were in operation. One of the company's principals was from Italy with experience in manufacturing sawing, bending, forming and processing. The company recently added new products to its portfolio. The range included sawing, forming, benders and measuring centres. Many of which were integrated with the sawing machines enabling cut pieces to be transferred automatically to flexible manufacturing systems (FMS). Computerized stock control formed an integral part of these systems. In addition to dealing in the sawing machines, the company was active in the tube industry by representing manufacturers of technologically advanced machines and systems.

Market Orientation of the Company

Following the interview with the executive, it was learned that the company supported the notion that market orientation was positively related to business performance of the company. On asking how market orientation was positively related to the performance of the company, the executive said that it was not simply that the company could make profit by selling machines in the first place but the company could make additional profit by providing quality service after sales, i.e. reliable support. So reliable customer support is the main criterion for the company to be financially viable. According to the executive there was no competition in the sector in which the company supplies machines to its customers. This meant that environmental factors, such as technology turbulence and competitive intensity were least applicable in the case. Although there was not much market turbulence in the market because of the unique specification of the machines, the executive believed that it was important to know what might be required in the future by the customers. The company stressed on the quality of its machine which led to retaining its customers. Customer retention rate for the company was about 80 per cent. Based on the interview and the analysis of the marketing materials of the company, some of the actions taken by the company to become more market oriented are presented in the following sections.

Reliable Customer Support

This company always took pride for its ability to service machine and its readiness to provide full customer support in the event of machine breakdown. To maintain the high level of customer support and service, the company held over one million spare parts and dispatched 93 per cent of orders received before noon on the same working day. Another aspect of customer support was taking care of potential customer's needs prior to sales. This company began by evaluating practically the buyer's needs, and continued with the offer of various financing packages, which included lease and lease/purchase schemes with the option of taking back the old equipment for a part exchange.

Once a machine has been installed a routine preventative maintenance was carried out on the machines. Such arrangement reduced maintenance costs in the long-term and helped customers get the best from their machines. On site operator training and applications engineering advice were also included in the package. If an emergency call-out was needed, a response was guaranteed within the hour. This company's engineers were located across the country; all had mobile phones and were quickly on site to get the equipment up and running again.

Value Addition

This company had a proclivity to supply value-added machines to its customers. This was important as the more value was added the greater were the mutual benefits, leading to a more secured customer base. For example, block saws were capable of cutting blocks consistently and accurately to size, saving time and money for both suppliers and customers. For the customers, value was added in that scrap was virtually eliminated. Added value was passed down the supply chain. Another example of value addition was the internal and external chamfering at both ends of accurately cut-to-length tube. Therefore, these machines were sufficiently flexible to cope with the large batch of production and capable of increasing output significantly.

From the customer's point of view, the benefits of getting this type of service from suppliers were numerous. For example, in harsh economic times the prospect of having value added operations completed by the supplier was particularly appealing. The value addition was appreciated to an extent that customers were prepared to pay extra for this service. Certainly, the company researched the market thoroughly to determine exactly what its customers needed in terms of accuracy and finished shape of the products and how best it could meet those needs.

Project Management

One of the reasons for the company's success over recent years had been its creation of a project management team to oversee the activities of the company's sales engineers. Under this management, at the stage of customer' enquiry, the team undertook an analysis to establish what the customer's real requirements were. Often, these were somewhat different from what the customers thought. In most cases, better alternatives were frequently found. Having arrived at the optimum solution, the next stage was to put forward a feasible plan for installation of the machines. Costs were calculated and discussions took place about how best to fund the investment. Once an agreement had been reached and an order was placed, the team carried out a detailed computer simulation of the system to be installed, using the customer's specific parameters to show how exactly the machine would work in practice. This was done not once, but continuously throughout the life of a project. The customer received regular updates and progress reports as the system developed. This commitment of the company to perfection and managing every aspect of an installation applied to all equipment it sold. One of the executives said, 'get that right, and the chances are that subsequent processes will be right first time as well.' Indeed, the company practised what was proposed by Kohli and Jaworski (1990): the use of information pertaining to customer's needs and the dissemination of the information inside the organization to be responsive to meet those identified needs.

Customers' Perceptions of the Market Orientation

The completed questionnaires were examined through gap analysis to test for the differences on market orientation between the supplier and its customers. The mean scores are reported in Table 7.5 on the next page.

Results and Discussions

Table 7.5 suggests that there are significant positive differences between the company and its customers on three statements: 5 (+2, entire business of the supplier can contribute towards creating a value for customers), 7 (+2.5, the supplier is prepared to disagree with the customer to achieve a better decision for them), and 10 (+ 1.5, the suppliers' representative from the manufacturing department interacts directly with us to assess how best they can serve us). This positive difference means that the company's expectation of market orientation is more than its customers' expectations, which may not be a desirable situation because it suggests that the company is over content with its level of market orientation. As demonstrated by the scores on the statement 5, although this company creates value for its customers, it is yet to be perceived by its customers. The statements 7 and 10 may not be applicable to this company as it acts as a technical agent for a principal company abroad, although the company makes modifications to its products to meet customer needs.

Table 7.5 Means for the company C and its customers

Items	Company A	Customers B	Differences A-B
1	2.00	5.00	-3.00*
2	4.00	5.50	-1.50*
3	3.00	3.50	-0.50
4	4.00	4.00	0.00
5	6.00	4.00	+2.00*
6	5.00	4.00	+1.00
7	6.00	3.50	+2.50*
8	6.00	5.50	+0.50
9	6.00	5.00	+1.00
10	5.00	3.50	+1.50*
11	2.00	2.50	-0.50
12	4.00	4.50	-0.50
13	5.00	4.00	+1.00
14	5.00	4.00	+1.00
15	5.00	4.00	+1.00
16	4.00	5.00	-1.00

On the contrary, a negative difference suggests that the customers' perceptions are higher than the company's perception of its market orientation. This may indicate that the company enjoys a good reputation, which makes the customers think good of the company. Negative scores were obtained on the statements 1 and 2, which are related to revealing customer needs through customer research, and offering products to customers that satisfy their needs, respectively. It suggests that the customers perceive more than company that the company does research into customers to reveal and satisfy their needs. This finding may be the outcome of the introduction of the project management team, which is being perceived positively by the customers. The finding offers support for the notion that being close to customers is paramount in discovering and satisfying their needs. As the project management team satisfies a further need of financially weak customers by offering them financial packages, the relationship with customers might have let them form good images of the company. In turn, this resulted in the customers' market orientation score for the company being higher than the company's own market orientation score.

Overall it appears that the company is perceived as market-oriented by its customers. In fact, customers perceive this company to be more market oriented than the company perceives. The difference of minus one on statement 16 confirms this perception. Supported by many prestigious customers, it is no surprise that the company has an unrivalled reputation in the UK. Their business has been built on supplying the very best equipment for production.

Company D

Established in 1948, the company designs and manufactures metal cutting machine tools. It was a private company whose major stocks were owned by the management team. It was located at four locations in the UK with a sales office in a Far Eastern country with an annual sales turnover above £0.7 m and 145 employees. Twenty-five percent of its tools were exported. As a part of being market orientated, it had a partnership with 31 public limited companies. The company had a specialized product range, such as single-point tool, resin bond wheels, vitrified bond wheels, metal bond wheels, electroplated products and polycrystalline diamond (PCD) tooling. For the production of the cutting tools, the company had different sites for different types of products. Table 7.6 shows the relationship between the products being produced and the industry being served. Sales contributions from different technologies are presented in Table 7.7. On the next page, Table 7.8 reports the application of products to the nature of industries.

Table 7.6 Products and industries

Products	*Industries*
Dressing tools	Bearing Manufacturing
Electropad	Carbide
Metal Bond	Decorative Glass
Polycrystaline Tooling	Electronics
Resin Bond	Flat Glass
Vitrified Bond	Optical Glass
Wear Parts	Precise Engineering

Table 7.7 Technologies and sales

Technology	*Sales*
Metal Bond	27%
Resin Bond	20%
Tooling	35%
Electroplated etc.	18%

Table 7.8 Industries and applications

Industry	*Applications*
Engineering	Motor vehicle
	Bearing
	Engine components
	Carbide
	Ceramics
Glass	Decorative
	Flat
	Optical
Oil exploration	Drill bits

Market Orientation of the Company

For the last few years, the company was running in loss because of mismanagement and a lack of customer focus. Before 1970s, the company had an excellent reputation for its innovative products but the success of the company took a toll because it failed to develop new products to match the competition and meet customers' needs. As a result, the company pursued a cost-cutting strategy to survive but ultimately started accruing heavy losses. In 1994, the Managing Director of the company bought the major portion of the company's shares to restart the company from scratch by reshuffling managers and by creating a marketing department. This department had the responsibility to conduct market research into the needs of the customers of machine tool cutters.

The prime task for the director was to rebuild the reputation for innovation. A philosophy of partnership - working with its customers to provide a total quality solution - was introduced to achieve customer satisfaction. It was thought that the satisfaction of their customer through the performance of the newly developed cutting tools was paramount for the business to be profitable. At the same time a market orientation was planned to be achieved by establishing a match between the company and its customers in terms of the pricing of the products. Further, it was realized that unless customers were sophisticated, they could not value the products in terms of quality and service, making difficult for the company to be market-oriented. Although the benefits of being market oriented cannot be denied, a technological orientation along with market orientation was the best possible combination for the company to be profitable in the cutting tool industry.

In order to differentiate from the competition, the company developed and launched a highly technologically improved machine tool cutter whose tip was made from Polycrystalline Cubic Boron Nitride (PCBN) and was the only drill of its kind in the world. This company had the patent for the product. A combination of design and technology proved to be very successful. As a result, the company was able to sell the PCBN tool at a premium price, leading to a profit. No doubt, the company had a vision for the technological orientation. In an effort to keep an eye on the future, the managers of the company talked to their customers about their requirements for the machine tools in the next few years. The company was also interested in knowing which machine tools cutters would be phased out by the customers in the future. Detailed action taken by the company is discussed in the following sections.

Partnership and Relationship

The company had oriented its business activities towards creating customer satisfaction through an enhanced relationship with its customer. Following the takeover by the new management team, it was necessary to build the brand image of the company. This relationship with customers appeared to be an appropriate strategy to get rid of the tarnished brand image of the company. According to Gronroos (1996) relationship marketing requires a shift of a corporate focus from transaction - oriented initiatives to value enhancing relationship activities. The usual tactical elements involved in relationship marketing are: direct contacts with customers and other stakeholders, a database containing pertinent information regarding customers, and a customer focused service system. In strategic terms, relationship marketing aims to redefine the business as a service business and the key competitive elements as service competition, when looking at the organization from a process management point of view and not from a functional department point of view. Certainly, the company recognized that the changes in the business philosophy were needed in providing customer satisfaction through relationship marketing.

Traditionally, the concept of marketing refers to operations where the mass marketing is manipulated through a marketing program consisting of product, price, place and promotion. Although the concept is not useless, the marketing mix may not fit well with certain competitive situations. The company was quick to realise that the mass marketing and the marketing mix approach might not lead the company to a superior performance while meeting the demands of its increasingly sophisticated customers. Therefore, the company transformed itself to be able to offer enhanced value around the core product and reliable service to its customers along with creating a trustworthy relationship.

Further, the company adopted an approach which organized its resources in the best possible way to satisfy the customers as soon as their needs were identified. To create customer care, resources, such as time, knowledge, personnel and technology were utilized. These were thought to be the core elements for the development of relationship marketing strategy. Adopting a relationship marketing strategy meant that the company should be totally customer oriented and should consistently work towards building long-term partnership with all their markets and that everyone in the company should work together to provide the best possible quality products and services.

Total Quality Management (TQM)

This company had unrivalled experience in the application of Diamond and other super abrasive materials in the manufacturing industry. The achievement of ISO

9002 was an indication of the company's commitment to producing high quality products. During the interview with the Managing Director, it was learned that focus on quality with an emphasis on relationship marketing was the philosophy of the management in creating a market-oriented company. This company understood that TQM was the application of quality principles for the integration of manufacturing processes within the organization. Therefore, it focused on TQM along with relationship marketing to achieve customer satisfaction and market orientation.

Investor in People

Even the best strategies are nothing more than a piece of paper if they are not implemented within a company. And the best way to put strategies into action is to invest in personnel because the biggest assets of a company are its trained personnel. Consistent with this thinking, the company developed a policy of investment in its employees to not only increase their productivity but also to educate them about customers and their needs. The importance of combining technology with customers was stressed upon to achieve market orientation. Particular attention was given to quality management to increase customer satisfaction. As a result, a concept total quality management (TQM) was introduced to the company. A major part of the program was to involve all employees and commit them to quality-enhancing activities. Providing employees with the necessary training was the main objective of the company. Apprentices and employees of the company were expected to be qualified to the highest standard. CAD/CAM operators, quality technicians, and members of the sales and marketing team completed both in-house training and attended recognized institutions outside the company to gain necessary qualifications. The company had a strong commitment to training and paid attention to it. Certainly, the right employee training, development and education at the right time provides big payoffs for the employer in increased productivity, knowledge, loyalty, and contribution.

Customers' Perception of the Market Orientation

The scores for the groups are reported in Table 7.9

Conclusions and Discussions

Table 7.9 suggests that the positive differences between the company and its customers are significant for statements 2, 7, 8, 9, 15 and 16: 2 (+2.33,the supplier offers products that are truly satisfying its customers' needs), 7 (+3.33, the supplier is prepared to disagree with its customers to achieve a better decision for them), 8 (+2.67, the supplier influences its customers by information rather than by pressure), 9 (+3.67, the supplier's representatives meet its customers once a year to assess their customers' needs), 15 (+ 2, the supplier takes corrective action immediately for any complaints), and 16 (+1.67, the extent to which the supplier is market oriented).

Table 7.9 Means for the company D and its customers

Items	Company A	Customers B	Differences A-B
1	4.00	4.00	0.00
2	6.00	3.67	+2.33*
3	6.00	5.00	+1.00*
4	2.00	4.00	-2.00*
5	5.00	4.33	+0.67
6	5.00	5.33	-0.33
7	7.00	3.67	+3.33*
8	7.00	4.33	+2.67*
9	7.00	3.33	+3.67*
10	3.00	2.33	+0.67
11	3.00	2.67	+0.33
12	3.00	4.67	-1.67*
13	3.00	4.67	-1.67*
14	4.00	4.67	-0.67
15	6.00	4.00	+2.00*
16	5.00	3.33	+1.67*

Although this company did not claim to be fully market oriented, it took the middle path between market and technological orientations. This notion is reflected in statement 7 (difference being +3.33) that the company is prepared to disagree with its customer to achieve a better decision for its customer. During the interview it was discovered that the company's director paid visit to the customer's site to help them achieve better solutions in terms of technology, quality and service. In one instance, the director opened a service station near the site of a customer in order to provide a high level of service on tooling. The company believed that in the cutting technology, most often customers were not aware of the capabilities of the new cutting tools. Therefore, a certain amount of disagreement was important to provide the best possible technical solution. Another difference in the means was observed in the statement 9 (difference being +3.67) that company's representatives did not visit periodically to assess the needs of their customers. As the company was in the process of restructuring and enhancing its reputation for being innovative, this finding was surprising. However the lack of resources meant that the company was not yet successful in establishing contacts more frequently with all its existing customers.

Finally, a difference of + 1.67 in the statement 16 suggests that the customers of the company perceive the company to be less market–oriented than does the company itself. Given the resources and the loss of reputation for innovation, the company is satisfied with its current perceptions, operations and profitability. This company is yet to recover fully from the losses but is expected to be profitable in the next couple of years.

Cross-case Analysis and Conclusions

Table 7.10 Comparison of companies

	Company A TO/LO, MO/HI	Company B TO/HI, MO/HI	Company C TO/LO, MO/LO	Company D TO/HI, MO/LO
TO	5.0	60.0	3.0	6.5
MO	5.8	6.2	3.6	5.3
S	87.0	91.0	72.0	76.0
B	69.3	74.0	67.5	64
S-B	17.7	17	4.5	12
% [*]	18.4	17.7	4.5	12.5

TO=Turnover in million pounds, MO=Market Orientation score on 7-point scale, S=Suppliers' score, B=Buyers' score, S-B=Difference

* Calculated as the ration of [S-B] and Scale range; range being 96 i.e. [16 * 7 - 16 * 1 = 96]; Mark-or = Market Orientation Score

To decipher the subtle similarities and differences, the 2 X 2 cell design was used to compare cases individually and simultaneously. The overall idea was to become familiar with each as a stand-alone entity. This process allowed the unique pattern of each case to emerge before researcher push to generalize the pattern across cases. In addition, the familiarity with individual cases accelerates cross-case comparison. In the following section I highlight the key findings for each case, followed by the cross-case conclusions.

Company A is a low turnover and a high market oriented company; its aim is to increase business performance as a direct result of achieving market orientation through the production of high quality user friendly CNC machines and through the export of these machines. This company is not fully successful in implementing its market orientation strategy as evidenced by a gap (18%) in the perceptions of market orientation between the company and its supplier. However, the difference may be attributed to the fact that the customers are yet to realize the benefits of the newly introduced machines. As the products are at the introduction stage of the product life cycle, the director of the company is optimistic that the customer would eventually rate the company as a high market oriented company.

Company B is a very high turnover and a high market oriented company. The gap between the company and its customers towards market orientation of the company is 17.7 per cent, which may be attributed to the fact that large companies find difficulties in coordinating their activities to become responsive towards their customers needs. Indeed the rate of responsiveness should vary according to the complexity of products and the market situations. It may also be noted that the company has undergone the merger of two companies. As a result, I would expect the company to be more market oriented. But the director of the company is content with the current level of market orientation and it has no intention to increase it. The director of company B cautions that it is not advisable to be fully market-oriented because exclusive attention to market can lead to the neglect of the existence of customer; there should be a combination of both market and technological orientations. The finding may suggest that performance of a company is more important than merely being perceived as a market orientated company by its customers.

Company C is a low turnover and a low market oriented company. This company's strategy to achieve market orientation is through providing a high level of customer support and value addition to its customers. Further, this company has adopted the concept of customer-oriented project management team whose prime task is to work with its customers and come up with a machine that is most suitable for the customers' needs. This team is also responsible for arranging financial packages should there be a need for it. Activities of this company towards the implementation of the market orientation are more or less perceived by its customers the same (difference being only 4.5%) as the company itself. This suggests that the company has the ability to meet and satisfy customer needs despite the fact that this company has a small annual sales turnover and limited resources. This could also mean that since the company is small, it is easier for the company to co-ordinate the activities among departments to be responsive to customer needs. The lean structure of the company may further contribute to achieving the market orientation. Of these four companies, this company is the most close to market orientation perceived by its customers.

Company D is a high turnover and a low market oriented company. To become a market-oriented company, it stresses the need for innovation, relationship marketing and organizational learning (by investing in employees) to create a superior value for its customers. This thinking is consistent with Hurley and Hull (1988) who highlight the requirement for innovation in marketing orientation, citing the dynamic nature of most markets which drives the need for innovation and re-orientation. Indeed, market orientation promotes organizational learning, and the organization's ability to learn then enhance performance. Further, the gap between the company and its customers on the perception of its market orientation is 12.5 per cent, an indication that the company's activities are aligned with the expectation of its customers. Given the limited resources and the new organizational structure, it appears that the company has an effective market oriented strategy.

Our cross-case conclusion is that market orientation has a different meaning for each different company. In each case the interpretation of the market orientation may be the correct for a particular company for a particular market. Further, as these companies represent different characteristics of an organization, there are differences (varying from 5 to 13 per cent) on perceptions of market orientation between suppliers and their customers. From the cross-analysis, it appears that all companies over rated their level of market orientation, that smaller companies were more market oriented than larger companies, and that not all companies are concerned about the customers' evaluation of their market orientation. Above all, the most important issue is the long-term profitability of the business, whether it is market, customer or technological orientation. Nonetheless, the study provides the support for the central tenet that market orientation leads to higher business performance, although the means to achieve market orientation may be different for different companies.

Therefore, I accept the proposition partially that the customer's perception of the market orientation is more important than a seller's own perception of market orientation in explaining the seller's own business performance. Although the acceptance of the hypothesis is not conclusive, it does shed light on the level of agreement on market orientation between suppliers and their buyers. The conclusion may be generalized into only theoretical prepositions as contribution to literature and to the general principles. However, these findings provide a basis for the development and testing the link between market orientation and business performance through the analysis of cases in other industries.

It is recognized that the gap analysis has a weakness in that the scores on the statements, which are ordinal, are assumed to proximate the property of interval scale (Birnbaum, 1992), which permits calculations. Nonetheless, Table 7.10 demonstrates the process that could be followed using interval data.

A future line of enquiry could be the case studies employing multiple levels of analysis within a single study (Yin, 1984), or employing multiple investigators to make the visits to case study sites in teams, which allows the case to be viewed from the different perspectives of multiple observers (Pettigrew, 1988), or multiple data collection to test hypotheses using triangulation methodology.

Given the importance of market orientation and its link to business performance, development of teaching cases (Quinn, 1980) would be of essence.

Chapter 8

Conclusions, Limitations and Future Research

Conclusions

Several insights can be drawn into the nature of market orientation based on the hypotheses formulated in the preceding chapters. Our objectives were to redevelop a market orientation scale, to identify underlying dimensions of market orientation, to examine the impact of these dimensions on business performance, and to detect variables, which could influence the market orientation-business performance relationship. Further, it was intended to test the hypothesis that the degree of market orientation varied respectively from highest to lowest according to the type of corporate culture as follows: market, adhocracy, clan, and hierarchical culture, and that the customer's perception of the market orientation was more important than suppliers' own perception of market orientation in explaining the suppliers' own business performance. Following sections report the results of the hypotheses.

The market orientation scale was developed by comparing the items between the market orientation scales of Narver and Slater (1990), Jaworski and Kohli (1993) and Deng and Dart (1994). Based on factor analysis, it was discovered that the market orientation scale, which is intended to be representative of set of market-oriented activities in the context of machine tool industry, was best represented by 24 items under four dimensions: customer focus, competitor focus, customer satisfaction focus, and marketing focus, having 10, 7, 5 and 2 items, respectively. Split-half method ensured the reliability of these dimensions. Some aspects of the validity of the scale were also performed.

To have further insight into the characteristics of these dimensions, the influence of each dimension on performance was assessed. Findings suggest that customer focus and customer satisfaction focus have a stronger impact on performance than the other dimensions. It was also revealed that competitor orientation has a U shape relationship with performance in the short-term but a positive linear relationship in the long-term.

The central tenet that market orientation is positively related to business performance is accepted. More specifically, results indicate that Market Orientation is significantly and positively related to Return on Investment, Sales Growth, Market Share, New Product Success, Customer Retention and Global Presence, but not significantly related to the overall performance of companies.

Next, the following hypotheses based on moderating effect of environments: that the lesser the extent of Market Turbulence, the greater the positive impact of Market Orientation on Return on Investment, that the lesser the extent of

Technological Turbulence, the greater the positive impact of Market Orientation on New Product Success, and that the higher the extent of Competitive Intensity, the greater the positive impact of Market Orientation on Sales Growth are supported.

Further, the hypothesis that the degree of Market Orientation varies respectively from highest to lowest according to the type of corporate culture as follows: Market, Adhocracy, Clan, and Hierarchical is partially supported. In the context of machine tool industry, the findings suggest that the degree of market orientation varied respectively from highest to lowest according to the following type of organizational culture: adhocracy, market, hierarchical and clan.

Finally, our results based on limited number of case studies partially suggest that the customer's perception of the market orientation may be more important than seller's own perception of market orientation in explaining the seller's own business performance. Therefore, a mixed support for the proposition was found.

Towards a Theory of Market Orientation in Marketing Literature

In the context of machine tool industry, the study contributes to the debate by redefining market orientation as market orientation as the set of activities coordinated in such a way that derives customer satisfaction through superior performance of products (machines) and related services while still being competitive in the marketplace. Redefining market orientation is necessary to reflect the role of customer satisfaction in addition to customer and competitions. We believe that only fulfilling customers need is not enough. To retain customers and to have repeat orders, customers must be satisfied. Satisfaction is prerequisite to market orientation. Satisfaction can only be derived through the superior performance of the current machines. Dissatisfaction is the result of unconfirmed expectations. Marketers who understand the impact of customer satisfaction on business performance will want to secure future sales order on the basis of the recommendations of currently satisfied end-users of the products because what happens in the current buying decision will affect future purchase decisions. Therefore, in the context of machine tool industry, we reflected the significance of customer satisfaction by redefining market orientation.

Further, we contribute to the theory by discovering that there are four distinct factors that best represent market orientation of machine tool companies. They are: customer focus, competitor focus, customer satisfaction focus and marketing focus.

Finally, the most significant and interesting contribution that emerged from the study is that low and high competitor focus activities contribute more to business performance than medium competitor focus activities. Although these differences are not significant but marginal in that the costs of becoming competitor oriented outweigh the benefits when the level of competitors' orientation is medium. These findings offer support for the view that companies become progressively less competitor oriented following the receipt of an order. This is because there is now more need to be customer oriented (to execute the order) than to be competitor oriented, as there are no further competitive activities till next competition for submission of tenders. As companies make progress through the execution of the orders, they tend to step up their competitive activities while still maintaining full focus on customer orientation in order to have a positive impact on performance. This trend is demonstrated by positive gradients for both customer and competitor orientations. It appears that companies do not lose focus of their customers at any time, but do adjust their level of competition-oriented activities given the resources they have to compete in the marketplace.

Implication for Managers

The implication for managers is that it pays to be customer oriented. They should develop a customer-oriented culture (e.g. keeping the whole business informed about major customers; products lines that are driven by market research; quick to modify products as per customer's need; identify the need of end users and interact frequently with other departments) before they endeavour to become competitor oriented. Competitor orientation should only be developed if companies are in possession of necessary resources to commit for investment and are ready to endure initial revenue losses. Our findings appear to indicate benefits for companies that have a medium to high competitor orientation. Clearly there is a need for cost-benefit analysis to be undertaken by managers before a competitor orientation strategy is pursued (e.g. assess the quality of existing products and services; collect industry information through informal means; seek opportunities to gain competitive advantages and getting marketing people involved with the product development team) as this brings profit only in the long term. With regard to customer satisfaction it is vital that a medium to high level of customer satisfaction is obtained by providing customers with custom made machines and high quality service after sales. This can be implemented by assessing the customer's product preferences, talking to end-users, agents, and distributors. Finally, marketing orientation, although the last dimension in terms of variance explained in the market orientation construct, is as important as the other dimensions. It implies that there is a need to communicate the attributes associated with the machines (either new or the improvement on the existing machines) to the customers in an effective way and that managers should be quick to respond to competitor campaigns targeted at their customer base.

Further, managers can use the market orientation scale by responding and scoring to the items on the scale at any point in time. Then a target should be set to achieve a higher score within a time frame on those items that needed improvement. The target can also be considered as a benchmark, which can be set by forming focus groups within the organization whose members are drawn from various departments, as one of the goals of being market-oriented is to have interfuncational co-ordination among departments, too. A repetitive use of the scale in future to measure the differences between the scores on the items and the scores on the bench-marked items will determine the extent to which managers are successful in achieving market orientation in the wake of market and technological turbulence.

I hope that the study gives food for thought to practising managers about how customer, competitor, customer satisfaction focus and marketing focus can contribute to enhancing performance of their companies in the short- and long-term in the light of external environmental variables. Finally, the findings offer an insight into the machine tool industry, but stop somewhat short of full generalizations. However, the comprehensive construct developed could and should be used as a test bed for further research into other manufacturing industry sectors.

Limitations

As with most research efforts, this study is not without limitations. It is recognized that the sample may not have same proportions of CNC and non-CNC companies as in the industry. Therefore, there is no claim that the sample is representative of the population; however, there is reasonable spread of companies based on number of employees and annual sales turnover.

The respondents were asked to score subjectively on a seven-point Likert scale. These evaluations are subject to personal bias and judgement errors. However, financial constraint necessitated the use of this methodology. Future research could include multiple-respondent technique and use objective data such as financial performance.

The study provides only a snapshot picture at a single point in time, thus the recommendations may be valid only if the external environmental variables are unaffected e.g. government regulations, foreign exchange, market growth, among others. It would be interesting to test whether these variables moderate the relationship between various dimensions of market orientation and business performance.

The modest sample size places limitations on the confidence in the findings. Repetition of the study with a bigger sample would help validate the findings. Nonetheless, the findings of the consequences of market orientation on performance do shed some light on understanding the impact of market-oriented activities.

The results are limited to those of association and not causation. Thus, a possible causal model can be built by employing a time series database.

Extension to the Study

As this study has discovered that market orientation has four significant underlying dimensions and that market orientation is positively related to business performance. The future study could consist of the development of a tool called MO-INDEX. The development of the tool can follow a methodology suggested by Phillips and Moutinho (1998). This tool can be used to measure the extent to which companies are market oriented. Then, the composite score of the MO-

INDEX can be related to the performance of the company to help diagnose the company's health. This managerial tool can aid machine tool manufacturers in evaluating the current effectiveness of their market orientation program.

Final Thoughts

Considering the study reported in this book, the main conclusions which can be drawn within the machine tool industry are as follows: there are four underlying factors of market orientation, namely, customer orientation, competitor orientation, customer satisfaction and marketing orientation. In addition, there are three secondary factors, namely, responsiveness, negligence and market information. In terms of relationships, the findings of this study reveal a significant association between market orientation and a variety of performance measures depending on the nature of the competitive environment. First, market orientation exerts a greater impact on return on investment when market turbulence is low. Second, market orientation has a greater effect on new product success when technological turbulence is low. Third, there is a stronger link between market orientation and sales growth when competitive intensity is high.

Generally, the findings lend support to the market orientation-performance hypothesis. However, much more work is required if we are to appreciate the mechanisms and implications of sound market oriented practices in the competitive contexts of both the machine tool industry and the wider national economy. It is important to explore further the dynamics of market orientation and of business performance in order to reinforce the universal significance of the marketing discipline. It is contended that a useful approach to this is to test the market orientation-performance paradigm based on a comprehensive framework. Given the global importance of markets and competitive success in today's marketplace, this issue will continue to attract the attention of marketing academics and practitioners alike. Indeed, understanding the organization of markets and behaviour of individuals within them remains a major challenge.

Evidently, important progress has been made but, undoubtedly, further theoretical work, as spelt out below, is still required. Returning to the key hypotheses of the marketing-performance paradigm advanced at the commencement of this thesis, while the studies reported in this thesis lay a foundation for analysing the relationship between market orientation and performance in the machine tool industry, further research efforts would add to existing knowledge by exploring new and related propositions. Future research could focus on processes for developing a broader market oriented culture and for taking advantage of it through reinforcing organizational processes, capabilities and strategies. Since in its highest form, a market orientation is conceptualised as a learning orientation (Day, 1994; Sinkula, 1994), future efforts should examine whether other variables such as entrepreneurial values, leadership style and organizational structure enhance the impact of a market orientation on performance. Some authors have set this agenda in motion (e.g., Morgan,

Katsikeas and Appiah-Adu, 1998) but more research is required in this area to capture the overall benefits of a learning organization's holistic approach to the marketing discipline. In this context, the following propositions could be examined:

* Market orientation will exert a stronger influence on competitive success when it is linked with entrepreneurial values such as tolerance for risk, proactiveness, and receptivity to innovation and active resistance to bureaucracy (Naman and Slevin, 1993).
* Market orientation will exert a greater effect on performance when it is coupled with facilitative leadership (Senge, 1990).
* Market orientation will have a higher impact on performance when it operates within a firm with an organic structure and flexible planning style (Slater and Narver, 1995).

Another area of recent research, which provides a stimulus for future research, is the issue of market orientation from an international marketing perspective. Due to the complexities of the global marketplace, practices, which epitomize a market orientation in a domestic context, may not represent this orientation in an international setting (Diamantopoulos and Cadogan, 1995). Given the arguments that market oriented firms tend to have more proactive international motives, and also tend to be more active and committed exporters than non market oriented firms (Dalgic, 1994), the following proposition could be tested in subsequent research: Market oriented organizations in international environments are more profitable in the international arena than non-market oriented firms.

A further area of challenge is to understand how a market orientation can be developed and maintained. How should these programmes be designed? Should the emphasis be essentially on cultural change, revamped work processes, organizational restructuring, new systems, revised incentives, or some other series of sound initiatives? This is a very important issue since managers would appreciate some insight into the characteristics of successful programmes for building market orientation. Such findings will offer guidance to executives on how to improve and redirect their firms' external orientation towards their markets.

Day (1994) combines strategic management theory with total quality management principles in an effort to suggest ways in which change programmes can be affected to enhance a market orientation. He suggests that firms become more market oriented by identifying and developing the unique capabilities that set market driven businesses apart, contending that such firms are superior in their market sensing, customer linking and channel bonding capabilities. Clearly, for these capabilities to be properly cultivated and fully exploited management needs to give simultaneous consideration to the values, beliefs and behaviours of individuals within the firm, augmented by changes in the organization structure, system, control, incentives and decision processes. A number of interesting future research avenues for marketing scholars arises from this thought provoking work. These include the need to explore the characteristics of change programmes that are productive in improving a market orientation and the factors that stimulate businesses to ardently strive for an enhancement in their orientation to the market. Moreover, can principles from other subject areas such as total quality management, human resources management and business re-engineering (Morgan and Piercy, 1992) be successfully adapted in managing change programmes designed to improve an organization's orientation to its market?

This thesis has argued that the utilisation of theoretical and empirical research offers a sound approach to investigating the effects of firm conduct or behaviour on performance, taking into consideration the intervening variables within the competitive environment. The central approach taken has been to focus attention on the competitive struggle between firms in the machine tool industry. There is a fundamental, endemic struggle for survival, growth and prosperity and striving to grow a business, grow a market and improve profits depends fundamentally on satisfying customer needs. Developing, adapting, modifying and enhancing the product or service, in order to secure competitive advantage is essential for survival and growth and the customer is at the heart of this enhancement of a firm's offering. In summary, the chapters in this thesis can be divided into four broad and distinctive types: an overview, clarification and elaboration of the existing literature; establishment of a framework for theory development; a well-defined empirical investigation; and detailed case studies.

The original contribution, it is hoped, can be seen in theory development and empirical investigation, a point reinforced in the articulation of the future direction for research in the machine tool industry.

Bibliography

Aaker, D.A. (1988), *Strategic Market Management*, John Wiley, New York.

Abbott, A. (1988), 'Workshop on Sequence Methods', *National Science Foundation Conference on Longitudinal Research Methods in Organizations*, (September).

Abell, D. F. (1980), *Defining the Business: The Starting Point of Strategic Planning*, Prentice Hall, Englewood.

Aiken, L. S. and West, S. (1991), *Multiple Regression: Testing and Interpreting Interaction*, Sage Publications, London.

Alan, M. (1998), 'P & G's New Horizon', *Compaign London*, March, pp. 34-35.

Albert, K. (1981), *Straight Talk About Small Business*, McGraw Hill, New York.

Alderson, W. (1957), *Marketing Behaviour and Executive Action*, Irwin, Homewood.

Althauser, R. P. (1971), 'Multicollinearity and Non-additive Regression Models', in Blalock, H. M. Jr. (ed.) *Causal Models in the Social Sciences,* Aldine-Atherton, Chicago, pp. 453-72.

Anderson, P. (1983), 'Decision Marking by Objection and the Cuban Missile Crisis', *Administrative Science Quarterly*, Vol. 28, pp. 201-222.

Appiah-Adu, K. (1997), 'Market Orientation and Business Performance: Do the Findings Established in Large Firms Hold in the Small Business Sector?', *Journal of EuroMarketing*, Vol. 6, pp. 1-26.

Appiah-Adu, K. (1998), 'Market Orientation and Performance: Empirical Tests in a Transition Economy', *Journal of Strategic Marketing*, Vol. 26, pp. 25-45.

Appiah-Adu, K. and Singh, S. (1998), 'Market Orientation and Performance: An Empirical Study of British SMEs', *Journal of Entrepreneurship*, Vol. 7(1), pp. 27-47.

Ames, B. C. (1970), 'Trapping Versus Substance in Industrial Marketing', *Harvard Business Review*, Vol. 48(July-Aug), pp. 93-102.

Armstrong, J. S. and Overton, T. S. (1977), 'Estimating Nonresponse Bias in Mail Survey', *Journal of Marketing Research*, Vol. 14(August), pp. 396-402.

Arndt, J. (1978), 'How Broad Should the Marketing Concept Be', *Journal of Marketing*, Vol. 42 (Jan), pp. 101-103.

Arnold, H. (1982), 'Moderator Variables: A Classification on Conceptual, Analytic and Psychometric Issues', *Organisational Behaviour and Human Performance*, Vol. 24, pp. 41-59.

Ashburn, A. (1993), '1992 Machine Tools Output: World Total Drops by $8 Billion', *American Machinist*, February, pp. 32-37.

Atuahene-Gima, K. (1995), 'An Exploratory Analysis of the Impact of Market Orientation on New Product Performance: A Contingency Approach', *Journal of Production, Innovation Management*, Vol. 12, pp. 275-93.

Atuahene-Gima, K. (1996), 'Market Orientation and Innovation', *Journal of Business Research*, Vol. 35, pp. 93-103.

Bagozzi, R. P. and Fornell, C. (1982), 'Theoretical Concepts, Measuring, and Meaning', in C. Fornell (ed.), *A Second Generation of Multivariate Analysis*, Vol. 2, Praeger, New York, pp. 24-38.

Balkrishnan, S. (1996), 'Benefits of Customer and Competitor Orientation in Industrial Markets', *Industrial Marketing Management*, Vol. 25, pp. 257-69.

Barksdale, H. C. and Darden, B. (1971), 'Marketers Attitude Towards the Marketing Concept', *Journal of Marketing*, Vol. 35(Oct), pp. 29-36.

Barnard, C. (1938), *The Functions of the Executive*, Harvard University Press, Cambridge.

Barnes, J. H., Daswar, A. K. and Gilbert, F. W. (1994), 'Number of Factors Obtained by Chance: A Situation study' in Wilson, E. J. and Black, W. (eds), *Development in Marketing Science*, Academy of Marketing Science, Nashville.

Barringer, B. R. and Bluedorn, A. C. (1999), 'The Relationship Between Corporate Entrepreneurship and Strategic Management', *Strategic Management Journal*, Vol. 20(5), pp. 421-44.

Bennett, P. D. and Harrel, G. D. (1975), 'The Role of Confidence in Understanding and Predicting Buyers' Attitudes and Purchase Intentions', *Journal of Consumer Research*, Vol. 2(Sept), pp. 110-17.

Bennett, R. and Cooper, R. (1981), 'Beyond the Marketing Concept', *Business Horizons*, Vol. 22(June), pp. 76-83.

Bentley, K. A. (1990), 'Discussion of the Link Between one Organisation Style and Structure and its Connection with its Market', *Journal of Product Innovation Management*, Vol. 7, pp. 19-34.

Bhargawa, M., Dubellarc, C. and Ramaswami, S. (1994), 'Reconciling Diverse Measures of Performance: A Conceptual Framework and Test of a Methodology', *Journal of Business Research*, Vol. 31, pp. 235-46.

Bhuian, S. N. (1996), 'Examining Market Orientation, its Antecedents and Consequences Among Saudi Manufacturing Companies', *American Marketing Association*, Summer, pp. 370-71.

Birnbaum, M. H. (1982), 'Controversies in Psychological Measurement', in Wegner, B. (ed.) *Social Attitudes and Psychological Measurement*, Lawrence Erlbaum, New York, pp. 41-86.

Blalock, H. M. (1979), *Social Statistics*, McGraw Hill, New York.

Block, B. (1989), 'Creating a Culture All Employees Can Accept', *Management Review*, (July), pp. 41-45.

Bohrnstedt, G. W. (1970), 'Measurement' in Wright, J. and Rossi, P. (eds) *Handbook of Survey Research*, Academic Press, New York.

Bollen, K. A. and Lennox, R. (1991), 'Conventional Wisdom on Measurement: A Structural Equation Perspective', *Psychological Bulletin*, Vol. 110, pp. 305-14.

Bolton, R. N. and Drew, J. H. (1991), 'A Longitudinal Analysis of the Impact of Service Changes on Customer Attitudes', *Journal of Marketing*, Vol. 55(Jan), pp. 1-9.

Borch, F. J. (1957), 'The Marketing Philosophy as a Way of Business Life', in Martin, E. and Garden, A. (eds) *the Marketing Concept, its Meaning to Management*, American Marketing Association, New York, pp. 3-16.

Bourgeois, L., Eisenhardt, K. (1988), 'Strategic Decision Processes in High Velocity Environments: Four cases in the Microcomputer Industry', *Management Science*, Vol. 34, pp. 816-35.

Brewer, J. and Hunter, A. (1989), *Multi Method Research: A Synthesis of Styles*, Sage Publication, California.

Bright, J. (1958), *Automation and Management*, Harvard Business School, Boston.

Busemeyer, J. R. and Jones, L. (1983), 'Analysis of Multiplicative Combination Rules: When the Causal Variables are Measured With Error?', *Psychological Bulletin*, Vol. 93, pp. 115-86.

Business Week (1950), 'Marketing Men Take Over in GE Units', June, pp. 30-36.

Buzzell, R. D. and Gale, T. B. (1987), *The PIMS Principle: Linkage of Strategy to Performance*, Free Press, New York.

Buzzell, R. D., Gale, T. B. and Sultan, G. M. (1975), 'Market Share: A Key to Profitability', *Haward Business Review*, Vol. 53(Jan-Feb), pp. 97-106.

Cadogan, J. W. and Diamantopoulos, A. (1995), 'Narver and Slater, Kohli and Jaworski and the Market Orientation Construct: Integration and Internationalisation', *Journal of Strategic Marketing*, Vol. 3, pp. 77-88.

Campbell, D. T. (1975), 'Degrees of Freedom and the Case Study', *Comparative Political Studies*, Vol. 8(July), pp. 178-193.

Cameron, K. and Freeman, S. (1991), 'Cultural Congruence, Strength, and Types: Relationship to Effectiveness', *Research in Organisational Change and Development*, Vol. 5, pp. 23-58.

Canning, G., Jr (1988), 'Is your Company Marketing Oriented', *Journal of Business Strategy*, Vol. 9(3), pp. 34-36.

Capon, N., Farley. U., and Hoening, S. (1990), 'Determinants of Financial Performance: A Meta Analysis', *Management Science*, Vol. 36(10), pp. 1143-59.

Caruana, A., Gauci, S. and Ferry, M. (1995), 'Market Orientation and Business Performance: Some Evidence From Malta', in Jobber et al. (eds) *Making Marketing Work*, pp. 123-31.

Cavusgil, S. T. and Zou, S. (1994), 'Marketing Strategy- Performance Relationship: An Investigation of the Empirical Link in Export Market Ventures', *Journal of Marketing*, Vol. 56, pp. 1-21.

Chakravarthy, B. S. (1986), 'Measuring Strategic Performance', *Strategic Management Journal*, Vol. 7, pp. 437-58.

Chamberlin, C. (1933), *The Theory of Monopolistic Competition*, Harvard University Press, Cambridge.

Chang, T. Z. and Chen, S. (1994), 'The Impact of a Market Orientation on Total Offering Quality and Business profitability', *American Marketing Association*, (Winter), p. 59.

Chee, L. K. and Peng, N. K. (1996), 'Customer Orientation and Buyer Satisfaction: The Malaysian Housing Market', *Asia Pacific Journal of Management*, Vol. 13(1), pp. 101-16.

Chow, G. C. (1960), 'Tests of Equality Between Sets of Coefficients in Two Linear Regression', *Econometrica*, Vol. 28(3), pp. 591-605.

Churchill, G. A., Jr (1979), 'A Paradigm for Developing Better Measures of Marketing Constructs', *Journal of Marketing Research*, (Feb), pp. 64-73.

Churchill, G. A., Jr. and Peter, J. P. (1984), 'Research Design Effects on the Reliability of Rating Scales: A Meta Analysis', *Journal of Marketing Research*, Vol. 21, (Nov), pp. 360-75.

Cohen, J. and Cohen, P. (1975), *Applied Multiple Regression for the Behavioural Sciences*, Lawrence Erlbaum, New Jersey.

Conduit, J. and Mavondo, F. (2001), 'How Critical is Internal Customer Orientation to Market Orientation', *Journal of Business Research*, Vol. 51, pp. 11-24.

Covin, J. G. (1991), 'Entrepreneurial Versus Conservative Firms: A Comparison of Strategies and Performance', *Journal of Management Studies*, Vol. 28(5), pp. 439-62.

Covin, J. and Slevin, D. (1989), 'Strategic Management of Small Firms in Hostile and Benign Environments', *Strategic Management Journal*, Vol. 10, pp. 75-87.

Cooper, R. G. (1979), 'The Dimensions of Industrial New Product Success and Failure', *Journal of Marketing*, Vol. 43(Summer), pp. 93-103.

Cooper, R. (1984), 'New Product Strategies: What Distinguishes the Top Performers', *Journal of Product Innovation Management*, Vol. 9, pp. 151-64.

Coopers, R. G. and Kleinschmidt, E. J. (1987), 'New Projects: What Separates Winners Fm Losers?', *Journal of Product Innovation Management*, Vol. 4, pp. 169-84.

Cravens, D. W., Holland, C. W., Lamb, C. W., Jr. and Montcrief, W. C. (1988), 'Marketing Role in Product and Service Quality', *Industrial Marketing Management*, Vol. 17, pp. 285-304.

Cravens, K. S. and Guilding C. (2000), 'Measuring Customer Focus: An Examination of the Relationship Between Market Orientation and Brand Valuation', *Journal of Strategic Marketing*, Vol. 8, pp. 27-45.

Crespi, I. (1961), 'Use of a Scaling Technique in Surveys', *Journal of Marketing*, (July), pp. 69- 72.

Cronbach, L. J. (1975), 'Beyond the Two Disciplines of Scientific Psychologist', *American Psychologist*, Vol. 30, pp. 116-27.

Cronin, J. J., and Taylor, S. A. (1992), 'Measuring Service Quality: A Re-examination and Extension,' *Journal of Marketing*, Vol. 56, pp. 55-68.

Crosier, K. (1975), 'What Exactly is Marketing?', *Quarterly Review of Marketing*, Vol. 1, pp. 21-25.

Dandridge, T., Mitroff, I. and Joyce, W. F. (1980), 'Organisational Symbolism: A Topic to Expand Organisational Analysis', *Academy of Management Review*, Vol. 5, pp. 77-82.

Davis, D., Morris, M. and Allen, J. (1991), 'Perceived Environmental Turbulence and its Effects on Selected Entrepreneurship and Organisational Characteristics in Industrial Firms', *Journal of Academy of Marketing Science*, Vol. 19, pp. 43-91.

Davis, D. D., Sharp, B. and Schlack, M. (1993), 'Can Digital Sustain Alpha's Edge?', *Datamation*, March, pp. 24-31.

Day, G. S. (1990), *Market Driven Strategy*, The Free Press, New York.

Day, G. S. (1994), 'The Capabilities of Market Driven Organisation', *Journal of Marketing*, Vol. 59(October), pp. 37-50.

Day, G. S. and Wensley, R. (1983), 'Marketing Theory With a Strategic Orientation', *Journal of Marketing*, Vol. 47(Fall), pp. 79-89.

Day, G. S. and Wensley, R. (1988), 'Assessing Advantage: A Framework for Diagnosing Competitive Superiority', *Journal of Marketing,* Vol. 52(April), pp. 1-20.

Deal, T. and Kennedy, A. (1982), *Corporate Culture*, Addison-Wesley, Reading.

Delbaere, M., Sivaramakrishnan, S. and Bruning, E. (2003), 'The Role of Knowledge Management in the Market Orientation-Business Performance Linkage', *Administrative Studies Association of Canada Conference*, Halifax, Nova Scotia, June 14-17, pp. 93-105.

Delgic, T. (1994), 'International Marketing and Market Orientation: An Early Conceptual Attempt at Integration', *Advances in International Marketing*, Vol. 6, pp. 69-82.

Deng, S. and Dart, J. (1994), 'Measuring Market Orientation: A Multi-factor, Multi-item Approach', *Journal of Marketing Management*, Vol. 10, pp. 725-42.

Deshpande, R., Farley, J. U., Webster, F. E., Jr. (1993), 'Corporate Culture, Customer Orientation, and Innovativeness in Japanese Firms: A Quadrad Analysis', *Journal of Marketing*, Vol. 57(January), pp. 23-27.

Deshpande, R. and Webster, F. E., Jr. (1989), 'Organisational Culture and Marketing: Defining the Research Agenda', *Journal of Marketing*, (January), pp. 3-15.

Dess, G. G. and Beard, D. W. (1984), 'Dimensions of Organisational Task Environments', *Administrative Science Quarterly*, Vol. 29(March), pp. 52-73.

Dess, G. G. and Davis, P. S. (1984), 'Porter's (1980) Generic Strategies as Determinants of Strategic Membership and Organisational Performance', *Academy of Management Journal*, Vol. 27, pp. 467-88.

Dess, G. G. and Robinson, R. B., Jr. (1984), 'Measuring Organisational Performance in the Absence of Objective Measure: The Case of a Privately Held Firm and Conglomerate Business Unit', *Strategic Management Journal*, Vol. 5, pp. 265-73.

Demirag, I. and Tylecote, A. (1992), 'The Effects of Organisational Culture, Structure and Market Expectations on Technological Innovation: A Hypothesis', *British Journal of Management*, Vol. 3, pp. 7-20.

Denzin, N. K. (1978), *Sociological Methods: A Source Book*, McGraw-Hill, New York.

Diamantopoulos, A. and Hart, S. (1993), 'Linking Market Orientation and Company Performance: Preliminary Evidence on Kohli and Jaworski's Framework', *Journal of Strategic Marketing*, Vol. 1, pp. 93-121.

Dickson, P. R. (1992), 'Toward a General Theory of Competitive Rationality', *Journal of Marketing*, Vol. 56(January), pp. 69-83.

Doyle, P., Saunders, J., and Wong, V. (1985), 'Comparative Investigation of Japanese and Marketing Strategies in the British Market', Report to ESRC (Ref: F200232034).

Doyle, P. and Wong, V. (1998), 'Marketing and Competitive Performance: An Empirical Study', *European Journal of Marketing*, Vol. 32 (5/6), pp. 514-35.

Drucker, P. (1954), *The Practice of Management*, Harper and Row, New York.

Dunn, M., Norburn, D. and Burley, S. (1994), 'The Impact of Organisational Values, Goals and Climate on Marketing Effectiveness', *Journal of Business Research*, Vol. 30, pp. 131-41.

Eccles, R. G. (1991), 'The Performance Measurement Manifesto', *Haward Business Review*, (January-February), pp. 131-37.

Edgett, S. and Thwaites, D. (1990), 'The Influence of Environmental Change on the Marketing Practice of Building Society', *European Journal of Marketing*, Vol. 24(12), pp. 35-47.

Eisenhardt, K., and Bourgeois, L. J. (1988), 'Politics of Strategic Decision Making in High Velocity Environments Toward a Mid-range Theory', *Academy of Management Journal*, Vol. 31, pp. 737-70.

Eisner, E. (1981), 'On the Differences Between Scientific and Artistic Approaches to Qualitative Research', *Education Researcher*, Vol. 10(4), pp. 5-9.

Farrell, O. C. and Lucas, G. H. (1987), 'An Evaluation of Progress in the Development of a Definition of Marketing', *Journal of Academy of Marketing Science*, Vol. 15, pp. 12-33.

Felton, A. P. (1959), 'Making the Marketing Concept Work', *Harvard Business Review*, Vol. 37(July-Aug), pp. 55-65.

Filion, F. L. (1975), 'Estimating Biases Due to Nonresponses in Mail Surveys', *Public Opinion Quarterly*, Vol. 40(Winter), pp. 482-92.

Filion, F. L. (1976), 'Explaining and Correcting for Nonresponse Biases Using Follows Ups of Non Respondent', *Pacific Sociological Reviews*, (July), pp. 401-408.

Filstead, W. J. (1979), 'Qualitative Methods: A Needed Perspective in Evaluation Research', in T. D. Cook and C. S. Charles (eds), *Qualitative and Quantitative methods in Evaluation Research*, Sage publication, California, pp. 33-48.

Financial Times (1994), *Ingenuity - The Financial Times Emerging Review*, March.

Financial Times (1996), *Machine Tool Sales Gain Rare Panache*, March.

Financial Times (1996), *Tool Sector Emerges with Sharper Act*, April.

Flippo, E. B. (1966), *Management: A Behavioural Approach*. Allyn and Bacon, Boston.

Fry, J. N. (1971), 'Personality Variables and Cigarette Brand Choice', *Journal of Marketing Research*, Vol. 8(August), pp. 298-304.

Gainer, B. and Pandanyi, P. (2002), 'Applying the Marketing Concept to Cultural Organization: An Empirical Study of the Relationship Between Market Orientation and Performance', *International Journal of Nonprofit and voluntary Sector Marketing*, Vol. 7(2), pp.182-93.

Geertz, C. (1973), *The Interpretation of Cultures*, Basic Books, New York.

Gerbing, D. W. and Anderson, J. C. (1988), 'An Updated paradigm for Scale Development Incorporating Unidimensionality and Its assessment', *Journal of Marketing Research*, Vol. 25, pp. 361-70.

Gersick, C. (1988), 'Time and Transition in Work Teams: Toward a New Model of Group Development', *Academy of Management Journal*, Vol. 31, pp. 9-41

Ghiselli, E. E. (1960), 'The Prediction of Predictability', *Educational and Psychological Measurement*, Vol. 20(Spring), pp. 3-8.

Ghiselli, E. E. (1963), 'Moderating Effects of Differential Reliability and Validity', *Journal of Applied Psychology*, Vol. 47(April), pp. 81-86.

Ghiselli, E. E. (1964), *Theory of Psychological Measurement*, McGraw Hill Company, New York.

Ghosh, B., Schoch, H., Huang, J., Lai, W., and Hooley, G. (1994), 'A Comparative Study of

Marketing Effectiveness: Profiles of Top Performers in Taiwan and Singapore', *Journal of International Marketing and Marketing Research*, Vol. 19(2).

Giaser, B. and Strauss, A. (1967), *The Discovery of Grounded Theory Strategies of Qualitative Research*, Wiedenfeld and Nicholson, London.

Gilmore, F. (1971), 'Formulating Strategy in Small Companies', *Harvard Business Review*, Vol. 47(3), pp. 71-83.

Goodenough, W. H. (1971), *Culture, Language and Society*, Addison-Wesley, Reading.

Golden, B. R. (1992), 'SBU Strategy and Performance the Moderating Effect of the Corporate-SBU Relationship', *Strategic Management Journal*, Vol. 13, pp. 145-58.

Gray, B., Matear, S. Boshoff, C. and Matheson, P. (1998), 'Developing a Better Measure of Market Orientation', *European Journal of Marketing*, Vol. 32(9/10), pp. 884-903.

Greenley, G. E. (1988), 'Managerial Opinions of Marketing Planning', *Omega*, Vol. 16(4), pp. 277-87.

Greenley, G. E. (1995), 'Market Orientation and Company Performance: Empirical Evidence From UK Companies', *British Journal of Management*, Vol. 6, pp. 1-13.

Grewal, R. and Tansuhaj, P. (2001), 'Building Organizational Capabilities for Managing Economic Crisis: The Role of Market Orientation and Strategic Flexibility', *Journal of Marketing*, Vol. 65(April), pp. 67-80.

Gronroos, C. (1989), 'Defining Marketing: A Market Oriented Approach', *European Journal of Marketing*, Vol. 23, pp. 52-60.

Gronroos, C. (1996), 'Relationship Marketing: Strategic and Tactical Implications', Management Decision, Vol. 34(3), pp. 5-15.

Guba, E. G. (1978), *Towards a Methodology of Naturalistic Inquiry in Educational Evaluation*. Centre for the Study of Evaluation, Los Angeles.

Guildford, J. P. (1954), *Psychometric Methods*, McGraw Hill, New York.

Gupta, A. K., Raj, S. P. and Wilemon, D. L. (1985), 'R and D and Marketing Dialogue in High Tech Firms', *Industrial Marketing Management*, Vol. 14, pp. 289-300.

Gupta, A. K., Raj, S. P. and Wilemon, D. L. (1986), 'A Model for Studying R and D in Marketing Interface in the Production Innovation Process', *Journal of Marketing*, Vol. 50, pp. 7-17.

Hallowell, R. (1996), 'The Relationship of Customer Satisfaction, Customer Loyalty and Profitability: An Empirical Study', *International Journal of Service Industry Management*. Vol. 7(4), pp. 27-42.

Hambrick, D. C. (1982), 'Environmental Scanning and Organizational Strategy', *Strategic Management Journal*, Vol. 3(2), pp. 159-74.

Hambrick, D. C. and Lei, D. (1995), 'Towards an Empirical Prioritization of Contingency Variables for Business Strategy', *Academy of Management Journal*, Vol. 28(4), pp. 763-88.

Harris, L. C. (1998), 'Cultural domination: The Key to Market-Oriented Culture?', *European Journal of Marketing*, Vol. 32(3/4), pp. 354-73.

Harris, S and Sutton, R. (1986), 'Functions of Parting Ceremonies in Dying Organizations', *Academy of Management Journal*, Vol. 29, pp. 5-30.

Hatti, J. (1985), 'Methodological Review: Assessing Unidimentionality of Tests and Items', *Applied Psychological Measurement*, Vol. 9, pp. 139-64.

Hatvany, N. and Pucik, C. V. (1981), 'Japanese Management Practices and Productivity', *Organisational Dynamics*, Vol. 9(4), pp. 5-21.

Hayes, R. A. and Abernathy, N. J. (1980), 'Managing Our Way to Economic Decline', *Harvard Business Review*, (July-August), pp. 67-77.

Hayes, R. A. and Wheelwright, S. C. (1984), *Restoring Our Competitive Edge*, Wiley and Sons, New York.

Hellievik, O. (1984), *Introduction to Causal Analysis*, Allyn, London.

Heskett, J., Loveman, T., Sasser, G., and Schlesinger, L. (1994), 'Putting the Service Profit Chain to Work', *Harvard Business Review*, (March-April), pp. 105-11.

Hise, R. T. (1965), 'Have Manufacturing Firms Adopted the Marketing Concept?', *Journal of Marketing*, (October), pp. 9-12.

Hobert, R. and Dunnette, M. D. (1967), 'Development of Moderator Variables to Enhance the Prediction of Managerial Effectiveness', *Journal of Applied Psychology*, Vol. 51(February), pp. 50-64.

Hofer, C. W. (1983), 'ROVA: A New Measure for Assessing Organisational Performance', *Advanced Strategic Management*, Vol. 2, pp. 43-55.

Hooley, G. and Lynch, J. (1985), 'Marketing Lessons From the UK's High Flying Companies', *Journal of Marketing Management*, Vol. 1, pp. 63-74.

Hooley, G., Lynch, J. and Shepherd, J. (1990), 'The Marketing Concept: Putting the Theory into Practice', *European Journal of Marketing*, Vol. 24(9), pp. 7-24.

Hooley, G. J., Lynch, J. E., Brooksbank, R. W. and Shepherd, J. (1988), 'Strategic Marketing Environment', *Journal of Marketing Management*, Vol. 4, pp. 131-47.

Hooley, G. J. and Newcomb, J. R., (1983), 'Ailing British Exports in Symptoms, Causes and Cures', *Quarterly Review of Marketing*, Vol. 8(4), pp. 15-22.

Horton, R. L. (1979), 'Some Relationships Between Personality and Consumer Decision Making', *Journal of Marketing Research*, Vol. 16(May), pp. 233-46.

Houston, F. S. (1986), 'The Marketing Concept: What it is and What it is Not', *Journal of Marketing*, Vol. 50(April), pp. 81-87.

Howard, J. A. (1983), 'Marketing Theory of Firms', *Journal of Marketing*, Vol. 47(Fall), pp. 90-100.

Huczynski, A. and Buchanan, D. (1991), *Organisational Behaviour - An Introductory Text*, Prentice Hall, New York.

Hunt, S. D. (1983), 'General Theories and the Fundamental Explanations of Marketing', *Journal of Marketing*, Vol. 47(Fall), pp. 9-17.

Hurley, R. F. and Hult, G. T. M. (1998), 'Innovation, Marketing Orientation and Organizational Learning: An Integration and Empirical Examination', *Journal of Marketing*, Vol. 62(3), pp. 42-54.

Innis, D. and Lalonde, B. (1994), 'Modelling the Effects of Customer Service Performance on Purchase Intentions in the Channel', *Journal of Marketing Theory and Practice*, (Spring), pp. 45-60.

Jaccard, J., Turrisi, R. and Wan, C. K. (1990), *Interaction Effects in Multiple Regression*, Sage Publication, London.

Jacobson, R. (1990), 'Unobserved Effects and Business Performance', *Marketing Science*, (Winter), pp. 74-85.

James, L., Demaree, R. and Wolf, G. (1984), 'Estimating Within Group Interrater Reliability With or Without Response Bias', *Journal of Applied Psychology*, Vol. 69(1), pp. 85-98.

Jauch, L. R. and Gluek, W. F. (1988), *Business Policy and Strategic Management*, McGraw Hill, New York.

Jaworski, B. J. and Kohli, A. K. (1993), 'Market Orientation: Antecedents and Consequences', *Journal of Marketing*, Vol. 57(July), pp. 53-70.

Jaworski, B. J., Kohli, A. K. and Kumar, A (1993), 'MARKOR: A Measure of Market Orientation', *Journal of Marketing Research*, Vol. 30(November), pp. 407-77.

Jaworski, B. J., Kohli, A. K. and Sahay, A. (2000), 'Market Driven Versus Driving Markets', *Journal of Academy of Marketing Science*, Vol. 28(1), pp. 119-35.

Jelenik, M., Smircich, L. and Hirsch, P. (1983), 'Introduction: A Code of Many Colours', *Administrative Science Quarterly*, Vol. 28, (September), pp. 331-38.

Jenning, P. and Beaver, G. (1985), 'The Managerial Dimension of Small Business Failure', *Journal of Strategic Change*, Vol. 4.

Jick, T. J. (1979), 'Mixing Qualitative and Quantitative Methods: Triangulation in Action', *Administrative Science Quarterly*, Vol. 24(December), pp. 602-11.

Johnson, P. and Gill, J. (1993), *Management Control and Organisational Behaviour*, Paul Chapman, London.

Joreskog, K. G. and Sorbom, D. (1988), *LISREL 7, A Guide to the Program and Application*, SPSS Inc, Chicago.

Jung, C. (1923), *Psychological Archetypes*, Routledge and Kegan Paul, London.

Kaiser, H. F. (1974), 'An Index of Factorial Simplicity', *Psychometrika*, Vol. 39, pp. 31-36.

Keith, R. J. (1960), 'The Marketing Revolution', *Journal of Marketing*, Vol. 24(January), pp. 35-38.

Kidder, T. (1981), *The Soul of a New Machine*, Little Brown, Boston.

Kidder, T. (1981), *The Soul of a New Machine*, Avon, New York.

Kiel, G., McVey, V. and McColl-Kennedy, J. (1986), 'Marketing Planning Practices in Australia', *Management Review*, (1).

Kirby, J. K. (1972), 'The Marketing Concept: Suitable Guide to Product Policy', *The Business Quality*, Vol. 27, pp. 31-35

Kirzner, I. M. (1979), *Perception Opportunity, and Profit*, University of Chicago Press, Chicago.

Kline, R. B. (1998), *Principles and Practices: SEM*, The Gulidford Press, New York.

Kohli, A. K. and Jaworski, B. J. (1990), 'Marketing Orientation: The Construct, Research and Managerial Implication', *Journal of Marketing*, Vol. 54(April), pp. 1-18.

Kohli, A. K., Jaworski, B. J. and Kumar, A. (1993), 'MARKOR: A Measure of Market Orientation', *Journal of Marketing Research*, Vol. 30(November), pp. 467-77.

Kohontek, H. K. (1988), 'Coupling Quality Assurance Programs to Marketing', *Industrial Marketing Management*, Vol. 17, pp. 177-88.

Konopa, L. J. and Calabro, P. J. (1971), 'Adoption of the Marketing Concept by Large Northeastern Ohio Manufacturers', *Akon Business and Economic Review*, Vol. 2(Spring), pp. 9-13.

Kotler, P. and Zaltman, G. (1971), 'Societal Marketing: An Approach to Planned Social Change', *Journal of Marketing*, Vol. 35, pp. 3-12.

Kotler, P. (1988), *Marketing Management: Analysis, Planning, Implementation and Control*, Prentice Hall, New Jersey.

Kotler, P. (1991), *Marketing Management Analysis, Planning and Control*, Prentice Hall, New Jersey.

Kotler, P. (1996) *Marketing Management, Analysis, Planning, Implementation and Control,* Prentice Hall, New Jersey.

Kotler, P. and Andreason, A. R. (1987) *Strategic Marketing for Nonprofit Organisations,* Prentice-Hall Inc, New Jersey.

Kotler, P. and Levy, S. J. (1969), 'Broadening the Concept of Marketing', *Journal of Marketing*, (January), pp. 10-15.

LAMBDAMAP (1987), *International Conferences on Laser Metrology and Machine Performance*, University of Huddersfield, Huddersfield.

Langerak, F., Nijssen, E., Droogenbroeck, S. and Delissen, J. (1996), 'Exploratory Results on the Influence of Market Orientation on the Competitive Advantage of Small and Medium Sized High Tech Firms in the Netherlands', *American Marketing Association*, (Summer).

Larreche, J. C. (1985), 'Quoted in Late Converts to the Group of Marketing and the Complexities that Often Lead to Failure', *Financial Times*, (April).

Lear, R. W. (1963), 'No Easy Road to Market Orientation', *Harvard Business Review*, Vol. 41, (September-October), pp. 53-60.

Levi-Strauss, C. (1963), *Structural Anthropology*, Basic Books, New York.

Levin, A. Y. and Minton, J. W. (1986), 'Determining Organisational Effectiveness: Another Look, and an Agenda for Research', *Management Science*, Vol. 32, pp. 514-53.

Levitt, T. (1960), 'Marketing Myopia', *Harvard Business Review*, Vol. 38(4), pp. 45-56.

Likert, R. (1967), *The Human Organisation: Its Management and Value*, McGraw Hill, New York.

Linchtenthal, J. D. and Wilson, D. T. (1992), 'Becoming Market Oriented', *Journal of Business Research*, Vol. 24, pp. 191-207.

Linchtenthal, J. D. and Beik, L. L. (1984), *A History of the definition of Marketing*, in J. N. Seth (ed.) *Research in Marketing*, Jai Press, Greenwich. pp. 133-63.

Liu, H. (1996), 'Market Orientation and Firm Size: An Empirical Examination in UK Firms', *European Journal of Marketing*, Vol. 29(1), pp. 57-71.

Liu, H. (1996), 'Pattern of Market Orientation in UK Manufacturing Companies', *Journal of European Marketing*, Vol. 5(2), pp. 77-100.

Loubser, S. S. (2000), 'The Relationship Between a Market Orientation and Financial Performance in South African Organizations', *Journal of Business Management*, Vol. 31(2), pp. 84-90.

Lucas, G. H., Jr. and Bush, A. J. (1988), 'The Marketing R and D Interface: Do Personality Factors Have Impact', *Journal of Product Innovation Management*, Vol. 5, pp. 257-68.

Lusch, R. F., Udell, J. G. and Laczniak, G. R. (1976), 'The Practice of Business', *Business Horizons*, Vol. 19(December), pp. 65-74.

Lusch, R. F. and Laczniak, G. R. (1987), 'The Evolving Marketing Concept, Competitive Intensity and Organisational Performance', *Journal of the Academy of Marketing Science*, Vol. 15(Fall), pp. 1-11.

Lumpkin, G. T. and Gregory G. D. (1996), 'Clarifying the Entrepreneurial Orientation Construct and Linking it to Performance', *Academy of Management Review*, Vol. 21(1), pp. 135-72.

MTTA (1995/96), *Machine Tool Technology Association, British Machine Tool and Equipment Directory 1995/96*, Berkshire.

MTTA (1998), *The World of Machine Tool: The Future of British Manufacturing*, Machine Tool Technology Association, London.

Magicnet (1996), *URL: http://www.magicnet.net/gencon/market.html.*

Malinowski, B. (1961), *Argonauts of the Western Pacific*, Routledge and Kegan Paul, London.

Manfred, W. (1984), *Handbook of Machine Tools, Metrological Analysis and Performance Tests*, Vol. 4, John Wiley & Sons, New York.

Marquandt, D. W. (1980), 'You Standardize the Predictor Variables in Your Regression Models, discussion of "A Critique of Some Ridge Regression Methods" by G. Smith and F. Campbell', *Journal of American Statistical Association*, Vol. 75, pp. 87-91.

Masiello, T. (1988), 'Developing Market Responsiveness Throughout Your Company', *Industrial Marketing Management*, Vol. 17, pp. 85-93.

Matsuno, K. and Mentzer, J. T. (2000), 'The Effects of Strategy Type on the Market Orientation-Performance Relationship', *Journal of Marketing*, Vol. 64(October), pp. 1-16.

Matsuno, K., Mentzer, J. T. and Ozsomer, A. (2002), 'The Effects of Entrepreneurial Proclivity and Market Orientation on Business Performance', *Journal of Marketing*, Vol. 66 (July), pp. 18-32.

Mayo, M. C. (1991), *Becoming Internationally Market Driven*, in T. L. Childers (ed.), *Marketing Theory and Applications*, American Marketing Association, Vol. 2, Chicago, pp. 454-60.

McArthur, A. W. and Nystrom, P. C. (1991), 'Environmental Dynamism, Complexity, and Munificence as Moderators of Strategy - Performance Relationships', *Journal of Business Research*, Vol. 23, pp. 349-61.

McCarthy, E. J. and Perreault, W. D., Jr. (1984), *Basic Marketing*, Richard D Irwin Inc, Homewood.

McDanial, S. and Parasuraman, A. (1986), 'Practical Guidelines for Small Business Marketing Research', *Journal of Small Business Management*, Vol. 24(1), pp. 1-8.

McKee, D. O., Varadarajan, P. R. and Pride, W. M. (1989), 'Strategic Adaptability and Firm Performance: A Market Contingency Perspective', *Journal of Marketing*, Vol. 53(July), pp. 21-35.

McKitterick, J. B. (1957) *What is the Marketing Management Concept*, in F. M. Bass (ed.) *The Frontiers of Marketing Thought and Science*, American Marketing Association, Chicago, pp. 71-92.

McNamara, C. P. (1972), 'The Present Status of the Marketing Concept', *Journal of Marketing*, Vol. 36(January), pp. 50-57.

Meziou, F. (1991), 'Areas of Strength and Weakness in the Adoption of the Marketing Concept by Small Manufacturing Firms', *Journal of Small Business Management*, (October), pp. 73-78.

Miles, M. (1979), 'Qualitative Data as an Attractive Nuisance: The Problem of Analysis', *Administrative Science Quarterly*, Vol. 24, pp. 590-601.

Miles, M. B. and Huberman, A. M. (1984), *Qualitative Data Analysis*, Sage Publication, Beverly Hills.

Miles, M. P. and Arnold, D. R. (1991), 'The Relationship Between Marketing Orientation and Entrepreneurial Orientation', *Entrepreneurship Theory and Practice*, (Summer), pp. 49-65.

Miller, C. and Cardinal, L. (1994), 'Strategic Planning and Firms Performances: A Synthesis of More Than Two Decades of Research', *Academy of Management Journal*, Vol. 37, pp. 1649-65.

Miller, D. (1987), 'The Structural and Environmental Correlates of Business Strategy', *Strategic Management Journal*, Vol. 8(1), pp. 55-76.

Miller, D. and Freisen, P. H. (1982), 'The Longitudinal Analysis of Organisations', *Management Science*, Vol. 28(9), pp. 1013-34.

Miller, D. and Freisen, P. H. (1983), 'Strategy Making and Environment: The Third Link', *Strategic Management Journal*, Vol. 4, pp. 221-35.

Miller, D. and Toulouse, J. (1986), 'Strategy Structure, CEO, Personality and Performance in Small Firm', *American Journal of Small Business*, Vol. 10(3), pp. 47-62.

Miller, L. O. and Leptos, C. B. (1987), 'Australia and the World: The Challenge', *Australian Institute of Management*, Melbourne.

Mintzberg, H. (1973), *The Nature of Managerial Work*, Harper and Row, New York.

Mintzberg, H. and McHugh, A. (1985), 'Strategy Formation in an Adhocracy', *Administrative Science Quarterly*, Vol. 30, pp. 160-97.

Morgan, R., Katsikeas, C. and Appiah-Adu, K. (1998), 'Market Orientation and Organisational Learning Capabilities', *Journal of Marketing Management*, Vol. 14, pp. 353-81.

Morgan, R. E. and Morgan, N. A. (1991), 'An Exploratory Study of Market Orientation in the UK Engineering Profession', *International Journal of Advertising*, Vol. 10, pp. 333-47.

Morgan, R. and Piercy, N. (1992), 'Market-led Quality', *Industrial Marketing Management*, Vol. 21, pp. 111-18.

Moriarty, R. T. and Thomas, J. K. (1989), 'High Tech Marketing: Concepts, Continuity, and Change', *Sloan Management Reviews*, (Summer), pp. 7-17.

Morrison, D. (1969), 'On the Interpretation of Discriminant Analysis', *Journal of Marketing Research*, Vol. 6(May), pp. 156-63.

Naidu, G. M. and Narayana, C. (1991), 'How Marketing Oriented are Hospitals in a Declining Market', *Journal of Health Care Marketing*, Vol. 11(1), pp. 23-30.

Naman, J. and Slevin, D. (1993), 'Entrepreneurship and Concept of Fit: A Model and Empirical

Test', *Strategic Management Journal*, Vol. 14, pp. 137-53.

Narver, J. C., Slater, F. S. and Tietje, B. (1998), 'Creating a Market Orientation', *Journal of Market-focused Management*, Vol. 2(3), pp. 241-55.

Narver, J. C. and Slater, S. N. (1990), 'The Effect of Market Orientation on Business Profitability', *Journal of Marketing*, (October), pp. 20-35.

NEDO (1982), 'Innovation in the UK', a NEDO report.

NMTBA (1981), *National Machine Tools Builders Association*, USA.

Neter, J., Wasserman, W. and Kutner, M. H. (1985), *Applied Statistical Models*, Irwin, Homewood.

Noble, C. H., Sinha, R. K. and Kumar, A. (2002), 'Market Orientation and Alternative Strategic Orientation: A Longitudinal Assessment of Performance Implication', *Journal of Marketing*, Vol. 66 (October), pp. 25-39.

Norburn, D., Birley, S., and Dunn, M. (1988), 'Strategic Marketing Effectiveness and its Relationship to Corporate Culture and Beliefs: A Cross-National Study', *International of Management Studies*, Vol. 18(2), pp. 451-68.

Norburn, D., Birley, S., Dunn, M. and Payne, A. (1990), 'A Four Nation Study of the Relationship Between Marketing Effectiveness, Corporate Culture, Corporate Values, and Market Orientation', *Journal of International Business Studies*, (Third Quarter), pp. 451-68.

Norusis, M. J. (1993), *SPSS for Windows Base System User's Guide*, SPSS, Inc, Chicago.

Nunnally, J. C. (1955) *Psychometric Testing*, McGraw Hill, New York.

Nunnally, J. C. (1964), *Educational Measurement and Evolution*, McGraw Hill, New York.

Nunnally, J. C. (1978, 88) *Psychometric Testing*, McGraw Hill, New York.

Olson, D. (1987), *When Consumer Firms Develop a Marketing Orientation* paper presented at Marketing Science Institute Miniconference on Developing a Marketing Orientation, (April), Cambridge.

OEF (1996), *Oxford Economic Forecasting*, 20th Annual MTTA Forecasting Seminar on November 05, 1996 at the Office of Cincinnati Milacron UK Ltd.

Oliver, R. L. and Swan, J. E. (1989), 'Equity and Disconfirmation Perceptions as Influences on Merchant and Product Satisfaction', *Journal of Consumer Research*, Vol. 16(December), pp. 372-83.

Orelwitz, A. (1993), 'Assessing the Relationship Between Market Orientation and Business Performance', *MBA Dissertation Submitted to the Faculty of Management, University of Witwatersrand*, South Africa.

Payne, A. (1988), 'Developing a Market-Oriented Organisation', *Business Horizons*, Vol. 31(May-June), pp. 46-53.

Parasuraman, A. (1980), 'The Impact of the Marketing Concept on New Product Planning', *Journal of Marketing*, Vol. 44(Winter), pp. 19-25.

Parasuraman, A., Zeithamal, V., and Berry, L. (1988), 'SERVQUAL: A Multi-item Scale for Measuring Consumer Perceptions of Service Quality', *Journal of Retailing*, Vol. 64, pp. 12- 40.

Parasuraman, A. and Deshpande, R. (1984), 'The Cultural Context of Marketing', *American Marketing Association Conference Proceedings*, Vol. 50, pp. 176-79.

Pearce, J. A. and David, F. R. (1987), 'Corporate Mission Statement: The Bottom Line', *Academy of Management Executive*, (May), pp. 109-16.

Pearce, J., Robinson, D. and Robinson, R. (1987), 'The impact of Grand Strategy and Planning on Financial Performance', *Strategic Management Journal*, Vol. 8, pp. 125-34.

Pelham, A. M. and Wilson, D. T. (1996), 'A Longitudinal Study of the Impact of Market Structure, Firm Structure, Strategy, and Market Orientation Culture on Dimensions of Small-Firm performance', *Journal of the Academy of Marketing Science*, Vol. 24(1), pp. 27-43.

Peter, J. P. (1979), 'Reliability in a Review of Psychometric Barrier and Recent Marketing

Practice', *Journal of Marketing Research*, Vol. 16(Feb), pp. 6-17.
Peters, T. J. and Austin, N. (1985), *A Passion for Excellence*, Random House, New York.
Peters, T. and Waterman, R. (1982), *In Search of Excellence: Lessons from America's Best-Run Companies*, Harper and Row, New York.
Peters, W. S. and Champoux, J. E. (1979), 'The Use of Moderated Regression in Job Redesign Decisions', *Decisions Sciences*, Vol. 10(January), pp. 85-95.
Pettigrew, A. (1988), 'Longitudinal Field Research on Change Theory and Practice', *National Science Foundation Conference on Longitudinal Research Methods in Organizations*, Austin.
Peterson, R. (1985), 'Creating Contexts for New Ventures in Stagnating Environments' in J. A. Hornaday, E. B. Shils, J. A. Timmons, H. K. Vesper (eds), *Frontiers of Entrepreneurship Research*, Babson Centre for Entrepreneurial Studies, Wesley.
Philips, L. W., Chang, D. R. and Buzzell, R. D. (1983), 'Product Quality, Cost Position and Business Performance: A Test of Some Key Hypotheses', *Journal of Marketing*, Vol. 47(Spring), pp. 26-45.
Phillips, P. A. and Moutinho, L. (1998), 'The MPI: A Tool for Measuring Strategic Marketing Effectiveness', *Journal of Travel and Tourism Marketing*, Vol. 7(3), pp. 41-59.
Piercy, N. and Morgan, N. (1994), 'The Marketing Planning Process: Behavioural Problems Compared to Analytical Techniques in Explaining Marketing Plan Credibility', *Journal of Business Research*, Vol. 29, pp. 167-78.
Pinfield, L. (1986), 'A Field Evaluation of Perspectives on Organizational Decision Making', *Administrative Science Quarterly*, Vol. 31, pp. 365-88.
Pitt, L., Caruana, A. and Berthon, P. R. (1996), 'Market Orientation and Business Performance: Some European Evidence', *International Marketing Review*, Vol. 13(1), pp. 5-18.
Pitt, L., Albert, C., Pierre, R. B. (1996), 'Market Orientation and Business Performance: Some European Evidence', *International Marketing Review*, Vol. 13(1), pp. 5-18.
Porter, M. (1980), *Competitive Strategy a Technique for Analysing Industries and Competitors*, The Free Press, New York.
Porter, M. (1985), *Competitive Advantage*, The Free Press, New York.
Power, C., Driscoll, L. and Bohn, E. (1992), 'Smart Selling', *Business Week*, Vol. 3(August), pp. 46-48.
Quinn, R. (1988), *Beyond Rational Management*, Jossey Bass Inc, San Francisco.
Quinn, J. B. (1980), *Strategies for Change*, Dow-Jones Irwin, Homewood.
Quinn, R. and Rohrbaugh, J. (1983), 'A Special Model of Effectiveness Criteria: Toward a Competing Values Approach to Organisational Analysis', *Management Science*, Vol. 29(3), pp. 363-77.
Radcliffe-Brown, A. (1952), *Structure and Function in Primitive Society*, Oxford University Press, London.
Reichardt, C. S. and Cook, T. D. (1979), *Beyond Qualitative versus Quantitative Methods*, in T. D. Cook and C. S. Reichardt (eds), *Qualitative and Quantitative Methods in Evaluation Research*, Sage Publications, Beverly Hills, pp. 7-32.
Reichheld, F. and Sasser, W. (1990), 'Zero Defections Comes to Services', *Harvard Business Review*, (September-October), pp. 105-11.
Roberts, F. B. (1990), 'Evolving Towards Product and Market Orientation: the Early Years of Technology Based Firms', *Journal of Product Innovation Management*, Vol. 7, pp. 274-87.
Robinson, R. (1982), 'The Importance of Outsiders in Small Firm Strategic Planning', *Academy of Management Journal*, Vol. 25, pp. 80-93.
Robinson, R. B. and Pearce, J. A. (1984), 'Research Thrusts in Small Firm Strategic Planning', *Academy of Management Review*, Vol. 9(1), pp. 128-37.

Robinson, R. and Pearce, J. (1988), 'Planned Patterns of Strategic Behaviour and their Relationship to Business Unit Performance', *Strategic Management Journal*, Vol. 9, pp. 43-60.

Rosenberg, M. (1968), *The Logic of Survey Analysis*, Basic Books, New York.

Rossman, G. B. and Wilson, B. L. (1985), 'Numbers and Words: Combining Quantitative and Qualitative Methods in a Single Large-Scale Evaluation Study', *Evaluation Review*, Vol. 9(5), pp. 627-43.

Ruekert, R. W. (1992), 'Developing a Market Orientation an Organisational Strategy Perspective', *International Journal of Marketing*, Vol. 9, pp. 225-45.

Rumelt, R. (1981), 'Evaluation of Strategy: Theory and Models', in D. Schendel and C. Hofer (eds) *Strategic Management: A New View of Business Policy and Planning*, Little Brown and Co, Boston, pp. 196-212.

Sachs, W. S. and Benson, G. (1978), 'Is it Not Time to Discard the Marketing Concept?', *Business Horizon*, Vol. 21(August), pp. 68-74.

Saunders, D. R. (1956), 'Moderator Variables in Prediction', *Educational and Psychological Measurement*, Vol. 16(Summer), pp. 209-22.

Saxe, R. and Weitz, B. A. (1982), 'The SOCO Scale: A Measure of the Customer Orientation of Sales People', *Journal of Marketing Research*, Vol. 19(August), pp. 343-51.

Schein, E. (1985), *Organisational Culture and Leadership*, Jossey-Bass, San Francisco.

Schein, E. (1990), 'Organisational Culture', *American Psychologist*, Vol. 45, pp. 109-19.

Schneider, B. and Arnon, R. E. (1983), 'On the Etiology of Climates', *Personnel Psychology*, Vol. 36(1), pp. 19-29.

Schneider, B. and Rentsch, J. (1988), 'Managing Climates and Cultures: A Futures Perspective', in J. Hage (ed.), *Futures of Organisations*, Lexington.

Schlegelmilch, B. B. and Ross, A. (1987), 'The Influence of Managerial Characteristic on Different Measures of Export Success', *Journal of Marketing Management*, Vol. 3(2), pp. 145-158.

Scherer, F. M. (1980), *Industrial Market Structure and Economics of Performance*, Rand McNally, Chicago.

Schoonhoven, C. B. (1981), 'Problems With Contingency Theory: Testing Assumptions Hidden Within the Language of Contingency Theory', *Administrative Science Quarterly*, Vol. 26(September), pp. 349-77.

Schwartz, H. and Davis, S. (1981), 'Matching Corporate Culture and Business Strategy', *Organizational Dynamics*, Vol. 10(Summer), pp. 30-48.

Senge, P. (1990), *Fifth Discipline*, Doubleday, New York.

Sexton, D. and Van-Auken, P. (1982), 'Prevalence of Strategic Planning in Small Business', *Journal of Small Business Management*, Vol. 20(3), pp. 20-26.

Shanklin, W. L. and Ryans, J. K., Jr, (1984), 'Organising for High-Tech Marketing', *Harvard Business Review*, Vol. 84, pp. 164-71.

Shapiro, B. P. (1988), 'What the Hell is Market Oriented?', *Harvard Business Review*, Vol. 88(6), pp. 119-25.

Sharma, S., Durand, R. M. and Gur-Arie, O. (1981), 'Identification and Analysis of Moderator Variables', *Journal of Marketing Research*, Vol. 18(August), pp. 291-300.

Shoham, A. and Ross, G. (1993), 'Export Performance: A Meta-Analytical Integration', in M. Levy and D. Grenal (eds), *Developments in Marketing Science*, Academy of Marketing Science, Vol. 16, Miami Beach, pp. 230-34.

Siguaw, J. A. (1994), 'The Influence of the Market Orientation of the Firm on the Sales Force Behaviour and Attitude', *Journal of Marketing Research*, Vol. 31(February), pp. 106-16.

Siguaw, J. A. and Diamantopoulos, A. (1995), 'Measuring market Orientation: Some Evidence on Narver and Slater's Three Component Scale', *Journal of Strategic Marketing*, Vol. 3, pp. 77-88.

Simon, H. (1991), *Why Sheer persistence is the Key to Japanese Success*, Financial Times, June.

Sin, L. Y., Tse, A. C., Yau, O. H., Lee, J. S., Chou, R. and Lau, L. (2000), 'Market Orientation and Business Performance: An Empirical Study in Mainland China', *Journal of Global Marketing*, Vol. 14(3), pp. 5-29.

Singh, S. (1998), 'A Study of the Relationship Between Market Orientation and Business Performance with Particular Reference to the Machine Tool Industry Based in the UK', *PhD Dissertation, Nottingham Trent University-Southampton Institute*, Southampton, England, UK.

Singh, S. and Ranchhod, A. (1998), 'Reassessing Market Orientation: An Empirical Test in the Machine Tool Industry', *British Academy of Management Conference*, (September) Nottingham, p. 16.

Singh, S. (2001), 'Is Corporate Culture-Business Performance Valid in the British Machine Tool Industry?', *30th European Marketing Academy Conference*, May 8-11, Bergen, Norway, pp. 13-20.

Singh, S. (2003), 'Effects of Transition Economy on the Market Orientation Business Performance Link: The Empirical Evidence from Indian Industrial Firms', *Journal of Global Marketing*, Vol. 6(4), pp. 73-96.

Singh, S. (2004), 'Impact of Staff Monitored Program on Firm's Performance: A Case of Angola', *Journal of African Business*, Vol. 5(1).

Singh, S. and Ranchhod, A. (2004), 'Market Orientation and Customer Satisfaction: Evidence From British Machine Tool Industry', *Industrial Marketing Management*, Vol. 33(2).

Sinkula, J. (1994), 'Market Information Processing and Organisational Learning', *Journal of Marketing*, Vol. 58, pp. 35-45.

Slater, S. F. and Narver, J. C. (1994), 'Does the Competitive Environment Moderate the Market Orientation Performance Relationship?', *Journal of Marketing*, Vol. 58(January), pp. 46-55.

Slater, S. F. and Narver, J. C. (1995), 'Market Orientation and the Learning Organisation', *Journal of Marketing* , Vol. 59(July), pp. 63-74.

Smircich, L. (1983), 'Concepts of Culture and Organisational Analysis', *Administrative Science Quarterly*, Vol. 28(September), pp. 339-58.

Smith, A. G. and Louise, K. S. (1982), 'Multi Method Policy Research: Issues and Applications', *American Behavioural Scientist*, Vol. 26(1), pp. 1-144.

Smith, J. K. and Heshusius, L. (1986), 'Closing Down the Conversation: The End of the Quantitative-Qualitative Debate', *Educational Researcher*, Vol. 15(1), pp. 4-13.

Snee, R. D. (1973), 'Some Aspects of Nonorthogonal Data Analysis, Part 1. Developing Prediction Equations', *Journal of Quality Technology*, Vol. 5, pp. 67-79.

Sohedi, A. W., Hart, S. and Tagg, S. (2001), 'Measuring Market Orientation in the Indonesian Retail Context', *Journal of Strategic Marketing*, Vol. 9, pp. 285-99.

Sounder, W. E. (1980), 'Disharmony Between R & D and Marketing', *Industrial Marketing Management*, Vol. 10, pp. 67-73.

Stewart, T. A. (1990), 'Do You Push Your People Too Hard?', *Fortune*, (October), pp. 121-28.

Sullivan, J. (1983), 'A Critique of Theory Z', *Academy of Management Review*, Vol. 8(1), pp. 132-42.

Sutton, R. and Callahan, A. (1987), 'The Stigma of Bankruptcy Spoiled Organizational Image and its Management', *Academy of Management Journal*, Vol. 30, pp. 405-36.

Szymanski, D. M, Bhardwaj, S. G. and Varadarajan, P. R. (1993), 'An Analysis of Market Share-Profitability Relationship', *Journal of Marketing*, Vol. 57(July), pp. 1-18.

Tauber, E. M. (1974), 'How Marketing Discourages Major New Projects?', *Business Horizons*, Vol. 22(3), pp. 22-26.

Tay, L. and Morgan, N. (2002), 'Antecedents and Consquences of Market Orientation in Chartered Surveying Firms', *Construction Management and Economics*, Vol. 20, pp. 331-41.

Taylor, J. W. (1992), 'Competitive Intelligence: A Status Report on US Business Practice', *Journal of Marketing Management*, Vol. 8(2), pp. 117-25.

Taylor, S. J. and Bogda, R. (1984), *Qualitative Research Methods: The Search for Meaning*, John Wiley, New York.

The Economist (1998), *Emerging Marketing Indicators*, May 7.

Thompson, J. D. (1967), *Organisations in Action*, McGraw-Hill, New York.

Thorn, R. (1996), *Industrial Marketing with Special Reference to Machine Tool Industry in the UK*, Advanced Manufacturing Technology Research Institute, Cheshire.

Turner, S. P. (1983), 'Studying Organisation Through Levi-Strauss' Structuralism' in G. Morgan (ed.), *Beyond Method: Social Research Strategies*, Sage Publications, Beverley Hills.

Turner, G. and Spencer, B. (1997), 'Understanding the Marketing Concept as an Organisational Culture', *European Journal of Marketing*, Vol. 31(2), pp. 110-21

Turrisi, R. and Jaccard, J. (2003), *Interaction Effects in Multiple Regression*, Sage Publications, London.

Van Maanen, J. (1988), *Tales of the Field On Writing Ethnography*, Chicago University Press, Chicago.

Vazquez, R., Santos, M. and Alvarez, L. (2001), 'Market Orientation, Innovation, and Competitive Strategies in Industrial Firms', *Journal of Strategic Marketing*, Vol. 9, pp. 69-90.

Venkatraman, N. (1989), 'Strategic Orientation of Business Enterprises: The Construct, Dimensionality and Measurement', *Management Science*, Vol. 35, pp. 942-62.

Venkatraman, N. and Prescott, J. (1990), 'Environment-Strategy Coalignment: an Empirical Test of its Performance Implications', *Strategy Management Journal*, Vol. 11, pp. 1-23.

Venkatraman, N. and Ramanujam, V. (1986), 'Measurement of Business Performance in Strategy Research: a Comparison approach', *Academy of Management Review*, Vol. 11, pp. 801-14.

Vekantraman, N. and Ramanujan, V. (1987), 'Measurement of Business Economic Performance: An Examination of Method Convergent', *Journal of Management*, Vol. 13(1), pp. 109-22.

Walker, O. C., Jr. and Ruekert, R. W. (1987), 'Marketing's Role in the Implementation of Business Strategies: A Critical Review and Conceptual Framework', *Journal of Marketing*, Vol. 51(July), pp. 15-33.

Webster, C. (1993), 'Refinement of the Marketing Culture Scale and the Relationship Between Marketing Culture and Profitability of a Service Firms', *Journal of Business Research*, Vol. 26, pp. 111-31.

Webster, C. (1995), 'Marketing Culture and Marketing Effectiveness in Service Firms', *Journal of Service Marketing*, Vol. 9(2), pp. 6-21.

Webster, F. E. (1981), 'Top Management's Concern About Marketing: Issues for the 1980s', *Journal of Marketing*, Vol. 45(Summer), pp. 9-16.

Webster, F. (1988), 'Rediscovering the Marketing Concept', *Business Horizon*, Vol. 31, pp. 29-39.

Webster, F. and Deshpande, R. (1990), *Analysing Corporate Culture in the Global Marketplace*, Marketing Science Institute, Cambridge.

Weick, K. E. (1979), *The Social Psychology of Organising*, Addison-Wesley, Reading.

Weitz, B., Sujan, H., and Sujan, M. (1986), 'Knowledge, Motivation and Adaptive Behaviour: A Framework for Improving Selling Effectiveness', *Journal of Marketing*, Vol. 50(October), pp. 174-91.

Wilkie, W. (1990), *Consumer Behaviour*, John-Wiley, New York.

Wilkins, A. and Ouchi, W. (1983), 'Efficient Cultures: Exploring the Relationship Between Culture and Organisational Performance', *Administrative Science Quarterly*, Vol. 28, pp. 468-81.

Wilson, D. T. and Ghingold, M. (1987), 'Linking R&D to Market Needs', *Industrial Marketing Management*, Vol. 16, pp. 207-14.

Wilson, K. J. (1992), 'The Role of the Sales Force in Creating And Sustaining Competitive Advantage in Industrial Markets, Marketing in the New Europe and Beyond', 1992 *Annual Conference Marketing Education Group, North West Centre for European Marketing*, Salford, pp. 468-78.

Workman, J. P., Jr. (1993), 'Marketing's Limited Role in New Product Development in One Computer Systems Firm', *Journal of Marketing Research*, Vol. 30, pp. 405-21.

Yi, Y. (1990), 'A Critical Review of Consumer Satisfaction', in V. Zeithaml (ed.) *Review of Marketing*, American Marketing Association, Chicago, pp. 68-123.

Yin, R. K. (1981a), 'The Case Study as a Serious Research Strategy', *Knowledge: Creation, Diffusion, Utilisation*, Vol. 3(September), pp. 97-114.

Yin, R. K. (1981b), 'The Case Study Crises: Some Answers', *Administrative Science Quarterly*, Vol. 26(March), pp. 58-65.

Yin, R. K. (1984), *Case Study Research: Design and Methodologies*, Sage Publications, Beverley Hills.

Zedeck, S. (1971), 'Problems With the Use of Moderator Variables', *Psychological Bulletin*, Vol. 76(October), pp. 295-310.

Index